T5-DHA-851

Hong Kong

A Study In Economic Freedom

Hong Kong
A Study In Economic Freedom

Alvin Rabushka
Hoover Institution on War, Revolution and Peace
Stanford University

The 1976–77 William H. Abbott Lectures in
International Business and Economics
The University of Chicago • Graduate School of Business

ISBN 0-918584-02-7

Contents

Preface and Acknowledgments

Shortly after the August 1976 meeting of the Mont Pelerin Society, held in St. Andrews to commemorate the 200th anniversary of the publication of Adam Smith's *The Wealth of Nations*, Dean Richard Rosett proposed that I deliver a series of lectures on the general subject of business and politics in Hong Kong. These were to be the 1976–1977 William H. Abbott Lectures in International Business and Economics, to be delivered at the Graduate School of Business of the University of Chicago. I accepted his invitation with great enthusiasm. There must be no better place than the University of Chicago at which to lecture on the subject of the industrial world's most robust example of a free-market economy. If Chicago symbolizes the concept of competitive markets, Hong Kong must surely reflect the living reality.

In this published volume, I have made a few slight changes to the original lectures that I delivered in April 1977. All tables, statistics and financial data have been updated to March 1978 and reflect the Financial Secretary's March 1978 Budget Address. The other major change consists of an expanded treatment of Hong Kong's money supply process. I have taken advantage of the knowledge of the fine monetary economists in residence at the Hoover Institution during the 1977–1978 academic year and for their comments thank Michael Darby of

U.C.L.A., Robert Barro of the University of Rochester, and Milton Friedman, who needs no identification.

I first went to Hong Kong in March 1963 as a Chinese language student and, though unintended, acquired a wife. Since then, I've made six additional trips to the colony. With the support of the National Fellows program of the Hoover Institution, I spent part of the summer of 1972 gathering materials to write *The Changing Face of Hong Kong*. This brief monograph highlighted the record of Hong Kong's remarkable economic growth and the structure of its free-market economy. A subsequent grant from the Hoover Institution enabled me to spend the fall of 1973 in Hong Kong conducting interviews, collecting published and unpublished statistics on budgetary allocations, and reading numerous government papers in order to write *Value For Money: The Hong Kong Budgetary Process*. This lengthier book analyzed policy-making and budgetary practice within the Hong Kong government.

The Abbott lecture series thus afforded a splendid opportunity to blend the two previous foci—one on market economics, one on political decision-making—into a comprehensive treatment of politics and business in Hong Kong, and add considerable new material as well. The four chapters that comprised the lectures now provide an introduction to anyone interested in the political economy of Hong Kong. They survey, in turn, (1) the political geography of Hong Kong and its historical evolution as a free market economy, (2) the political status of the colony to show why public officials are able to conduct responsible fiscal, economic and monetary policies, (3) a businessman's perspective on the economic freedom in Hong Kong, and (4) Hong Kong's future prospects. I have also used this lecture series to make an initial statement on the general subject of economic freedom, which I have researched for the past three years. Is Hong Kong unique or, an example of many cases of ecomomic freedom? Under what circumstances is the competitive market economy likely to emerge?

Both past and present support provided by the Hoover Institution has been invaluable in the preparation of these lectures. It is enormously helpful to have distinguished monetary economists as colleagues. It is equally helpful to have as colleagues a number of social scientists who are skilled analysts of the market economy. The presence of these scholars creates an environment that is especially conducive to thinking and writing about the subject of economic freedom.

I have many people to thank. For helpful comments on various parts of the manuscript, I thank Terry Anderson of Montana State University, Hugh O. Nourse of the University of Missouri-St. Louis, Allan H. Meltzer of Carnegie-Mellon University, and Dan Usher of Queen's University. Sir John Cowperthwaite, former Financial Secretary of Hong Kong [1961–1971], cleared up a number of mistakes in the preparation of my first two studies of Hong Kong and, in the process, kept me from wayward paths in the preparation of this volume. I also want to thank Philip Haddon-Cave, Financial Secretary of the Hong Kong government. Not only have I made use of his affectionate phrase "automatic corrective mechanism," but he has faithfully supplied me with his speeches, budget addresses, and annual economic reports over the past few years; without them I could not have prepared these lectures.

The first draft of these lectures was typed by Samirah Taha and the final version by Ilse Dignam; I thank them both. For shepherding the lectures through publication, my thanks to Betty McGuire.

Finally, I want to thank Dean Rosett for extending the invitation to deliver the 1976–1977 Abbott Lectures, despite the blizzard conditions of early April. But I also want to thank his entire family for their splendid hospitality during my two visits to Chicago in April 1977. No family has been a more gracious host.

Alvin Rabushka
Stanford, California
April 1978

The Evolution of a Free Society

I.

I can think of no more fitting place at which to discuss the political economy of Hong Kong than the University of Chicago. Many informed lay persons see Chicago as the spiritual and intellectual home of free-market economics. Economists, as the limited survey I've conducted indicates, identify Chicago with attempts to use notions of price theory and competition in the marketplace to explain a wide variety of real world economic behavior. Economists and non-economists alike suspect that the Chicago economist is a nineteenth century economic liberal—a devotee of laissez-faire—who disguises himself in a twentieth century analytical cloak labelled "positive economics." Right or wrong, I share this suspicion and undertake this pilgrimage to honor the modern-day descendants of Adam Smith. The tribute I offer is an exposition of the industrial world's most robust example of a truly competitive marketplace—the British Crown Colony of Hong Kong. This fascinating Gladstonian dream world stands in sharp contrast with today's increasingly interventionist, regulated and heavily taxed world. The people of Hong Kong live their economic lives free from both government's heavy hand of interference in their private affairs and what have so often become inquisitorial systems of high rates of taxation.

I must confess a certain measure of disappointment that Hong Kong receives so little analysis from the advocates of market economics. To date, the lion's share of printed words emanates from the typewriters of British expatriate intellectuals who condemn Hong Kong for its stubborn refusal to imitate the errors and declining fortunes of the mother country. These collectivist critics of Hong Kong's market economy regard making money and having low taxes as inherently immoral. One can, I think, count the number of American economists who study Hong Kong's political economy on the fingers of one hand, or at most two. I hope in this lecture series to kindle interest in Hong Kong's government and economy, which so closely approximate the classical notions of economic freedom and limited government. Hong Kong may be the one place where government consciously restricts its economic activities.

The Market Economy

I would like to begin this tale of one city by describing for you what is affectionately and appropriately termed the "automatic corrective mechanism."[1] This corrective mechanism continuously alters internal costs and prices to bring them quickly into line with costs and prices in the rest of the world. The flexible movement of internal costs and prices, with associated changes in output and employment, brings about internal and external equilibrium at all levels of world trade, and maximum economic growth. It also dictates the central elements of fiscal, economic, and monetary policy.

The automatic corrective mechanism takes its cue from the fact of Hong Kong's almost complete internal resourcelessness and hence heavy dependence on external trade. Hong Kong must import 85 percent of its food, nearly all industrial raw materials, and all its capital equipment. The colony pays this import bill primarily by exporting 90 percent of its manufactured goods; capital imports and the sale of invisible services also helps.

Net external transactions tie in directly to the internal cost/price structure. How does this process work? Suppose demand for Hong Kong products grows rapidly. Rapid growth would sharply increase the demand for labor and drive up costs and prices in the short run. This would tend to increase the rate of growth of imports (because people have more money to spend) and lower the rate of growth of exports (because Hong Kong's prices would rise relative to those of its competitors). The economy would automatically tend to deflate, thereby slowing down the rise in internal costs and prices and stimulating the production of goods for export at more competitive prices. The process repeats continuously.

Hong Kong's experience between 1973 and 1976 illustrates the automatic corrective mechanism adjusting through one full cycle of boom, recession, and recovery. Hong Kong's boom reflected a peak of activity in the world economy in 1973, when real per capita gross domestic product in the territory increased 10 percent. Real wages in the colony peaked in March 1973. Boom, however, gave way to recession. The previous year had witnessed a domestically generated monetary squeeze. A large net inflow of funds through the foreign exchange market in 1971 and 1972, associated with investment on the stock market sharply increased the money supply and rapidly drove up property prices and rents. An equally rapid decline in stock market prices—the Hang Seng Index declined from a high of 1,775 to a low of 160 in just a matter of months—and the consequent outflow of foreign exchange resulted in a sharp deceleration in the growth of the money supply.

But the full impact on the economy did not appear until the oil price increase late in 1973 diminished world economic activity and overseas demand for Hong Kong products. Reductions in output and employment in the manufacturing sector followed, which in turn reduced demand for imports. Deflationary forces appeared most strongly in the labor market. Unemployment rose to 9.1 percent by September 1975, a figure 2.3 percentage points higher than in 1971. Real wage rates also declined substantially in the two years from 1973 to 1975. (The index of real daily wage rates in the industrial sector fell at an annual rate of 9.7 percent from March 1973 to March 1975.)

The sharp recessionary forces of late 1973 through 1975 induced a severe squeeze on costs and prices, which led to a rapid and marked recovery in Hong Kong's competitive position in overseas markets. The elimination of inflation in Hong Kong's economy was accompanied by a very large inflow of funds from overseas, yielding a surplus in the overall balance of payments. The effect was to increase the assets of the banking system and leave it liquid and able to expand lending as demand recovered. Consumer prices increased only 1 percent in 1975 compared with 14 percent in 1971 and 18 percent in 1973.

Throughout 1976 the economy recovered briskly. Domestic exports increased 51 percent and imports rose 39 percent in the first six months of 1976 compared to the first six months of 1975. The volume of domestic exports grew 36 percent and imports 37 percent in the same period. The September 1975 unemployment rate of 9.1 percent fell to 5.6 percent by March 1976. Manufacturing employment rose by 18 percent between March 1975 and March 1976. This improvement took place as the Consumer Index increased only 2.8 percent over the first seven months of 1976, against the first seven months of the previous year. For the whole of 1976, real gross domestic product increased by an estimated 18 percent. By the end of 1976, real wages had almost returned to the peak level of March 1973, and have since steadily risen. Indeed, the number of manufacturing establishments increased 25 percent between September 1975 and September 1976, probably some sort of world record.

Why was this economic correction so efficient? A first reason is that the economy is free to respond to external price movements. In the absence of tariffs and other restrictions on the movement of goods and services, external price movements are quickly transmitted to the internal cost-price structure, with the consequent effect of increased or decreased external competitiveness. Furthermore, wages move freely up and down as determined by demand and supply; they are not rigidly held by collective bargaining arrangements to levels above the market clearing price.

Second, labor and capital are both highly mobile in Hong Kong's geographically compact, light industrial economy. The labor force is mobile between industries and trades; rigid craft demarcation does not prevail, nor is entry protected by trade unions. This mobility makes for an efficient labor force. Similarly, the preponderance of small-scale light industries operating in like premises means that capital can be switched from the making of one product to another, or from one industry to another, depending on external market demand.

The government's economic, fiscal, and monetary policies play an integral role in the workings of the automatic corrective mechanism. We can best describe economic policy as one of non-interventionism. The practice of non-interventionism by the Hong Kong government is symmetrical: freedom from bureaucratic controls means that the private sector must accept the consequences of its own mistakes. The treasury does not bail out firms that cannot survive the test of market competition. Shareholders must accept the risk of losses and bankruptcy against the prospective benefits of rising share prices. Dependence on external trade means that the market disciplines of bankruptcy and unemployment must not be hampered by the government. Indeed, dependence on external trade makes it unwise for government to interfere to encourage or discourage any particular economic activity, since businessmen with intimate knowledge of overseas markets must make decisions as to which industries to develop and in which markets to sell. Finally, Hong Kong's externally dependent economy must be left free to adjust to world prices, and this fact reduces considerably the scope for internal management of the economy. Action to maintain the level of aggregate demand and real incomes in Hong Kong's open economy would be quickly self-defeating.

The corrective mechanism is automatic, in part, because of the practice of responsible fiscal policy. Public expenditure is not allowed to grow faster than the economy as a whole, for economic growth can only be sustained if it is export-led. If public expenditure were to outpace growth in domestic output for any length of time, then resources would be diverted from the private sector and impair the longer-term growth prospects of the economy. Moreover, resources are typically used less profitably in the public than in the private sector, and in the process the growth rate of the economy is lowered. This shift in resource use leads, in turn, to higher internal costs and prices, balance of payments difficulties, and a still slower growth rate, meaning a lower standard of living.

Fiscal policies must, where possible, be neutral in their impact on the cost/price structure and investment decisions. Legitimate public spending must be distinguished from misconceived public expenditure, and the citizenry must be told that the government cannot use the fiscal system for unreasonable social purposes, except at great cost to the standard of living.

Finally, the term "automatic corrective mechanism" denotes the relationship between Hong Kong's money supply and the net balance of payments position, and is a particularly apt way to characterize an open economy. An overall surplus or deficit in the balance of payments must lead either to an appreciation or depreciation of the exchange value of the Hong Kong dollar or to an increase or decrease in the domestic money supply because of an increase or decrease in the liquid assets of the banking system. Banks in Hong Kong, as elsewhere, generate multiple credit expansion on the basis of their liquid assets.

The exchange rate of the Hong Kong dollar is affected by the various transactions that, taken together, make up the balance of payments. These include the balance of merchandise trade, the sale of such invisibles as tourism, insurance, shipping, and financial services, and the movement of capital funds. If Hong Kong's inflation is less than elsewhere, overseas investors will shift assets to Hong Kong. Because capital funds flow freely and quickly in and out of Hong Kong—Hong Kong is a completely free market in money—the Hong Kong price level quickly adjusts to world price levels through changes in the Hong Kong dollar exchange rate.

It is official policy in Hong Kong to let market forces determine the Hong Kong dollar exchange rate, though government sometimes intervenes to smooth out short-term erratic movements. But to say that the corrective mechanism is *automatic* means that budgetary and fiscal means are not used as a tool of monetary policy. Taking one year with another, financial authorities strive for a balance between revenue and expenditure. Thus, fiscal policy is both neutral and predictable insofar as the money supply and the value of the Hong Kong dollar are concerned.

Hong Kong avoids budget deficits and debt finance in the belief that deficit financing leads inevitably to an increase in the domestic money supply, largely to prevent interest rates from rising. In the case of Hong Kong's externally dependent economy, an increase in the domestic money supply would push up domestic costs and prices, causing a consequent decline in exports and growth of imports. Although the adjustment process would bring internal costs and prices back into line with world prices and ultimately restore the colony's overseas competitiveness, in the short run, resources would be misallocated and temporary damage in the form of adjustment costs borne by the trading sectors would be done to the efficiency of the economy. So, any attempt by Hong Kong to spend its way out of a recession would, by raising internal costs and prices, induce imports and reduce overseas competitiveness. Since the government is not prepared to maintain a fixed exchange rate well above or below that set in the market, we can say that the relationship between the balance of payments and the money supply is automatic, because the internal cost/price structure adjusts, through changes in the exchange rate, to eliminate any persistent surplus or deficit in external accounts.

By now you have recognized that this "automatic corrective mechanism" is simply the market mechanism, which rests on a keystone of individual responsibility and individual ownership. Neither should there be artificial barriers to individual ownership, nor punitive restrictions on the acquisition of wealth. The strong work ethic that accompanies individual initiative must not be eroded by inflation and redistributive fiscal policies. Market economies may impose short-term adjustments, but produce in the longer term a tendency towards rising real incomes, full employment, and a minimum of internally generated inflation. Lags in the adjustments to external forces may lengthen with increasing sophistication of the economy, but the basic logic of the adjustment mechanism remains incontrovertible. The requirement that Hong Kong's business community respond to the free play of market forces means that the internal cost/price structure will be flexible both downwards and upwards and thereby maximize economic growth.

Hong Kong is a living laboratory in which to observe marketplace competition; more than any other industrial society, it depends on market forces to allocate resources. How can we explain this twentieth century adherence to a nineteenth century economic world view? To answer this question we must look into many facets of Hong Kong's past and present: its people and its government, its resources and its geography, its business attitudes and regulations, and its utterly fascinating political status. These subjects take up the first three chapters; the fourth will entail a bit of crystal ball gazing along with a preliminary comparative exploration into the Hong Kongs of yesteryear to learn why competitive market economies are so rare.

Three circumstances provide the incentives and constraints which allow Hong Kong's market economy to function relatively free from government intervention. We have already seen the extent of the colony's dependence on external trade. The balance of this chapter will focus on relevant aspects of political geography; the next addresses the constitutional and political status of colonial government. Let us begin by setting the context with some background on the colony and its people.

The Colony and Its People

The British Crown Colony of Hong Kong lies inside the tropics on the southeast coast of China, adjoining the province of Kwangtung.[2] It consists of a small part of the Chinese mainland and a scattering of offshore islands, the most important of which is Hong Kong Island. The twin cities of Victoria (on Hong Kong Island) and Kowloon (on the peninsula directly facing Victoria), which overlook Hong Kong's magnificent natural harbor, are about ninety miles southeast of Canton and forty miles east of Portuguese Macao. Including recent reclamations, the total land area of the colony is 404 square miles, of which Hong Kong Island and a number of small adjacent islands comprise 29.2 square miles. Kowloon and Stonecutter's Island comprise another 4.3 square miles. The New Territories, which consist of part of the mainland and more than 230 offshore islands, have a total area of 370.5 square miles.

Of the 404 square miles in the colony, only 12 percent is used for farming; 76 percent is marginal unproductive land. Within the remaining 12 percent (about 48 square miles), most of Hong Kong's people live and work in sixteen thousand acres (twenty-five square miles) of extensively built-up areas. The 1961 census revealed that 80 percent of Hong Kong's more than 3 million residents live in approximately eight thousand acres (thirteen square miles), of which half was residential and a tenth was industrial. Hong Kong is one of the most densely populated places in the world. But the overall density figure of 4,187 per square kilometer at the end of 1975 includes a wide variety of densities in individual areas. Mong Kok, situated in Kowloon, is the most densely populated district, with 154,677 per square kilometer (or 400,613 per square mile). The overall metropolitan density is 17,098 per square kilometer; and for the New Territories, 468.[3]

More than 4.5 million people now live within Hong Kong's 404 square miles. The population increased from 3.6 million in 1965 to just under 4.4 million in 1975: an increase of 21 percent. Over 98 percent of the population can be described as Chinese on the basis of language and place of origin; of these, 58 percent were born and have lived their whole lives in Hong Kong. Of the 69,000 non-Chinese residents, 62 percent originate from Commonwealth countries outside Hong Kong and consist chiefly of British, Indian, Australian, Singaporean, and Canadian subjects. Many of these work in the civil service, or do

Table 1: Estimated Total Population, 1945–1977

Mid-year	*Population*	*Mid-year*	*Population*
1945	600,000[a]	1962	3,305,200
1947	1,750,000	1963	3,420,900
1948	1,800,000	1964	3,504,600
1949	1,857,000	1965	3,597,900
1950	2,237,000	1966	3,629,900
1951	2,015,300	1967	3,722,800
1952	2,125,900	1968	3,802,700
1953	2,242,200	1969	3,863,900
1954	2,364,900	1970	3,959,000
1955	2,490,000	1971	4,045,300
1956	2,614,600	1972[b]	4,103,500
1957	2,736,300	1973[b]	4,219,300
1958	2,854,100	1974[b]	4,345,200
1959	2,967,400	1975[b]	4,379,900
1960	3,075,300	1976[b]	4,444,000
1961	3,168,100	1977[b]	4,514,000

Note: [a]Rough estimate
[b]End of year

Sources: *Hong Kong Statistics 1947–1967* (Hong Kong: Census & Statistics Department, 1969), Table 2.2, *Hong Kong Monthly Digest of Statistics* (Hong Kong: Census & Statistics Department), April 1972, Table 2.5, and *Hong Kong Annual Reports* for the Years 1972–1975, and *1978 Edition, Hong Kong in Figures.*

business in Hong Kong. Non-Commonwealth persons include over six thousand Americans and small communities of Japanese, Filipinos, Indonesians, Portuguese, Germans, Dutch, Koreans, and French.

Throughout its history, Hong Kong has willingly accepted political refugees. Several hundred thousand entered Hong Kong in the 1930s during the Sino-Japanese War. Later, however, the Japanese forcibly deported large numbers of Chinese to China during the World War II occupation of Hong Kong, in order to ease the local food problems. Four years of Japanese rule reduced the population of Hong Kong to about one-third of its prewar size of 1.8 million. But after Hong Kong's liberation from the Japanese in 1945, the deported Chinese returned at a rate approaching one hundred thousand a month. By the end of 1947, the population had regained its prewar size. Still another influx took place during the Chinese civil war of 1948 and 1949, when nearly a half million people, mainly from Kwangtung province, Shanghai, and other commercial centers, entered the colony. Thus, aside from returning residents, immigrants increased the population between 1945 and 1956 by approximately one million, of which some seven hundred thousand were refugees.

Resources

The small colony of Hong Kong is almost entirely lacking in natural resources.[4] Its mineral wealth is negligible, consisting of a modest amount of iron ore, building stone, kaolin clay, graphite, lead, and wolfram. Most of the activity in the mining and quarrying industry concentrates on the production of building stone and sand. As is shown in Table 2, only one-seventh of its land is arable; the colony cannot, therefore, feed itself, and throughout most of its history has encountered difficulty in maintaining an adequate water supply. Virtually all the materials of industry, and the vast majority of foodstuffs, must be imported.

Table 2: Land Utilization in Hong Kong

Class	*Area (Square miles)*	*Percentage of Whole*	*Remarks*
Built-up (Urban Areas)	49.0	12.1	Includes roads and railways.
Woodlands	49.3	12.2	Natural and established woodlands.
Grass and scrub lands	237.3	58.6	Natural grass and scrub.
Badlands	16.8	4.1	Stripped of cover. Granite country. Capable of regeneration.
Swamp and mangrove lands	4.8	1.2	Capable of reclamation.
Arable	42.0	10.4	Includes orchards and market gardens.
Fish ponds	5.6	1.4	Fresh and brackish water fish farming.

Source: *Hong Kong 1976. Report for the Year 1975*, p. 47.

Hong Kong sits on the edge of an eroded mountain chain which extends along the south coast of China. Consisting mainly of sedimentary and granitic rocks, the hilly topography restricts extensive agricultural activity. In general, the shortage of land has always been one of Hong Kong's chief concerns. The island, upon its occupation by the British in 1841, had scarcely any flat land. Reporting on this condition, *The Times* (London), candidly remarked on 17 December 1844:

> "The place has nothing to recommend it, if we except the excellent harbor. The site of the new town of Victoria . . . is most objectionable, there being scarcely level ground enough for the requisite buildings, and

the high hills, which overhang the locality, shut out the southerly winds and render the place exceedingly hot, close and uncomfortable."[5]

The colony's masters undertook reclamation within ten years of its first occupation in 1841. Reclamation generally involves two steps. Spoil removed from quarrying and cutting back into the hills, which is an essential preparation for much of Hong Kong's building, is dumped into one of the surrounding bays; the building space is thereby extended simultaneously in two directions. The first praya reclamation scheme was completed in 1851, followed by further projects in 1867, 1884, 1885, and 1891.

Reclamation continued throughout the next eighty years; the largest schemes, totaling 415 acres, were completed around 1930. Since the war, 170 acres have been reclaimed for the Kai Tak airport, about 588 acres at Kwun Tong for private, commercial, industrial, and residential development, and an additional 500 acres at Kwai Chung. Massive new land formation exceeding 1,700 acres is in progress for the development of new towns in the New Territories.

Water shortages, as mentioned, have also posed severe difficulties. There are no rivers and only a few large streams in the colony. In consequence, satisfying the colony's water needs depends upon collecting and storing rainwater by systems of catchwaters and reservoirs, the bulk of which are located in the New Territories. This problem is complicated by the fact that the rainy season is concentrated in the five summer months from May to September; within these five months sufficient water must be impounded to last throughout the winter.

Between 1841 and 1863, Hong Kong Island was served by wells and hillside streams; a similar mode supplied Kowloon from 1860 until 1910. But dependence on these sources proved risky and inadequate. Accordingly, in 1864 a reservoir was built on the island at Pok Fu Lam but it, too, soon proved inadequate. In 1889 the completion of the Tai Tam scheme increased the total storage to 312 million gallons, which was subsequently raised to 511 million gallons with the completion of the Wong Nai Chung scheme in 1899.

Across the harbor, the first reservoir was in service in Kowloon in 1910, and the opening of the Shek Li Pui reservoir in 1922 provided Kowloon with a capacity of 468 million gallons. Additional schemes in Aberdeen, the Peak, and Kowloon steadily expanded the colony's storage capacity. By World War II, the colony had thirteen reservoirs that together could store 5,970 million gallons.

But the increasing demand for water rendered the prewar supply insufficient, leading, on occasion, to water rationing. At times, the mains would be opened for only three to four hours a day, or worse, for a four-hour period once every four days, as in 1963, during my first extended stay in Hong Kong. These conditions motivated construction of the Tai Lam Chung reservoir in the New Territories, which went into service in 1957 with a storage capacity of 4,500 million gallons; of Shek Pik on Lantau in 1963 with 5,390 million gallons; and of Plover Cove in the New Territories in 1969 with approximately 50,000 million gallons. Since 1968, Hong Kong has maintained a continuous water supply. To insure a steady flow for the future, the High Island scheme has commenced in the New Territories. It is scheduled for completion in 1979 with a capacity of 50,000 to 75,000 million gallons. A 40-million-gallons a day desalter is available for use in years of low rainfall. Some of the colony's flushing and fire-fighting systems have been converted to salt water, and most new buildings are installed with salt-water flushing systems. Salt water composed 16 percent of overall 1975 usage. Should a new shortage develop, the colony could by mutual agreement with China increase its purchase of fresh water, the annual quota now being set at 18,500 million gallons. In 1975, for example, China piped an extra 2,500 million gallons, and in 1977 agreed to an extra 6,000 million gallons.

Although water needs seem adequately provided for, Hong Kong still has an acute lack of suitable sites and premises for industry, and no local coal, oil, or other source of power. Nor is the colony's internal market sufficiently large to offer a solid base for economic expansion. In addition, there has been no tariff wall or other protectionist device to shield the growth of its flourishing industry from overseas competition.

An Economic History: 1841–Present

Hong Kong became a British possession in 1841 for the simple purpose of trade with China. Hong Kong prospered as an entrepôt free port, a mart and storehouse for goods in transit to Asia and the West. This entrepôt activity gradually diminished after World War II when the transition to industrial economy took place.

In keeping with Hong Kong's entrepôt character, the first industrial venture was shipbuilding and repairing. A number of ancillary industries were then established to cater for the seafaring trade: a large graving dock, a sugar refinery, a rope factory, and other service activities. Up until World War I the economic structure of the colony was of a number of enterprises linked with the operation of the port, with few cottage industries. Hong Kong became the headquarters of the major merchant houses trading in the Far East. As a clearing house of trade between the East and the rest of the world, it soon developed specialized associated services such as banking, insurance, accountancy, and legal services. It also developed such special facilities as the gold market. After World War I, the area of the entrepôt trade grew to cover much of the Far East and Southeast Asia.

Manufacturing industries are today the mainstay of Hong Kong's economy. The economy is to a large extent dependent upon export-oriented light manufacturing industries and a myriad of servicing industries operating within a free port, free enterprise environment. Highly developed banking, insurance, and shipping systems developed in the entrepôt era have facilitated and promoted the development of manufacturing industries. In its free port tradition, Hong Kong has no tariffs and few restrictions on the import of commercial goods. The establishment of Hong Kong as a vigorous industrial power has taken place without recourse to outside economic assistance, despite formidable obstacles arising from political circumstances beyond local control.

Hong Kong's post–World War II transformation from a trading to an industrial economy was so dramatic that between 1938 and 1956 it had successfully absorbed a doubling of its population. In the wake of the war, a good many of the factors that contribute to rapid industrial development were attained in Hong Kong almost overnight. The Communist revolution in China produced a massive influx of refugees, but most importantly it also resulted in Shanghainese capital and entrepreneurial skill moving to the safe haven of Hong Kong. Hong Kong's industrial revolution is thus attributable to the three resources brought with the Chinese political refugees: labor, new industrial techniques from the North, and new capital seeking employment and security. Additional circumstances favoring investment in industry included the relatively high adaptability and diligence of this labor force, the weak trade union organization, the lack of legislation fixing minimum wages and limiting working hours, and the extremely low taxation on business profits.

Hong Kong's remarkable economic record is clearly revealed in its trade statistics. Although official total balance of payments figures are not available, the value of merchandise trade is reported. Table 3 shows that domestic exports increased at an annual average rate of 13.8 percent between 1961 and 1971. In the four years 1968–1971, the annual growth in the value of domestic exports was 26, 25, 17, and 11 percent, respectively, and was as high as 44 percent in 1976 following a recession. It is likely that industrial production grew at an annual average rate of at least 30 percent between 1950 and 1964. As a result, income figures rose faster in Hong Kong than elsewhere in Asia. It was generally accepted in 1959 that Hong Kong's industrial wages were the third highest in Asia, after Japan and Singapore. Industrial wages are now generally believed to be higher only in Japan, having doubled between 1961 and 1971, a net increase of 50 percent in average real wages. In some industries, real wages more than doubled.

Imports have also risen at a dramatic rate; their 1977 value of HK$48,701 million was 175 percent over 1970. Foodstuffs consumed about 17 percent of all imports, with the balance consisting mainly of raw materials and semi-manufactured goods for industry, capital goods such as machinery and transport equipment, and consumer goods for both local consumption and the tourist trade. A good portion of these imports is subsequently re-exported; the value of this entrepôt trade totalled HK$6,973 million in 1971, an increase of 140 percent over 1970. All of these figures have risen dramatically up through 1977.

The large and growing trade deficit, which reached HK$3,868 million in 1977, is in part attributable to invisible receipts from tourism which account for about five percent of gross domestic product; the earnings from ser-

Table 3: Value of Merchandise Trade (in HK$ Millions)

Year	*Imports*	*Exports*	*Re-exports*	*Total Trade*	*Trade Balance*
1947	1,550	1,217		2,767	−333
1948	2,077	1,583		3,660	−494
1949	2,750	2,319		5,069	−431
1950	3,788	3,715		7,503	−73
1951	4,870	4,433		9,303	−437
1952	3,779	2,899		6,678	−880
1953	3,872	2,734	Included in Exports	6,606	−1,138
1954	3,435	2,417		5,852	−1,018
1955	3,719	2,534		6,253	−1,185
1956	4,566	3,219		7,776	−1,356
1957	5,150	3,016		8,166	−2,134
1958	4,594	2,989		7,583	−1,605
1959	4,949	2,282	996	8,227	−1,671
1960	5,864	2,867	1,070	9,801	−1,927
1961	5,970	2,939	991	9,900	−2,040
1962	6,657	3,318	1,070	11,045	−2,269
1963	7,412	3,831	1,160	12,403	−2,421
1964	8,550	4,428	1,356	14,334	−2,766
1965	8,965	5,027	1,502	15,494	−2,436
1966	10,097	5,730	1,833	17,660	−2,534
1967	10,449	6,700	2,081	19,230	−1,668
1968	12,472	8,428	2,142	23,042	−1,902
1969	14,893	10,518	2,679	28,090	−1,696
1970	17,607	12,346	2,892	32,845	−2,369
1971	20,256	13,749	3,414	37,420	−3,093
1972	21,764	15,245	4,154	41,164	−2,364
1973	29,005	19,474	6,525	55,004	−3,005
1974	34,120	22,911	7,124	64,156	−4,084
1975	33,472	22,859	6,973	63,304	−3,640
1976	43,293	32,629	8,920	84,850	−1,736
1977	48,701	35,004	9,829	93,534	−3,868

Sources: *Hong Kong Statistics 1947–1967*, Table 6.1; *Hong Kong Monthly Digest of Statistics*, April 1972, Table 5.1; and *The 1978–79 Budget: Economic Background*, Table 15.

vices of shipping, insurance and banking; and very substantial capital inflows. All make possible a substantial excess of imports over exports.

The colony did not publish official estimates of national income until the printing of the 1973–74 Budget Estimates, which contain an estimate of GNP for 1967–1971. Now national income estimates have been extrapolated back to 1961. These estimates show that over the period 1966–76, total gross domestic product at current market prices increased at an average annual rate of 15.6 percent, or 7.8 percent in real terms. Per capita gross domestic product at current market prices increased at an average annual rate of 13.3 percent, or 5.7 percent in real terms. Performance of the economy between 1961 and 1966 exceeded even these percentages.

Earlier data that exist on the gross national product or on overall growth rates in national income are based on a number of scholarly estimates which enjoy, perhaps, somewhat less reliability. K. R. Chou provides the following overall average growth rates per year in the gross national product: 1948–49, 22.6 percent; 1950–54, 5.3 percent; 1955–59, 9.0 percent; and 1960–64, 13.6 percent.[6] Szczepanik estimates the annual rates of income growth for the seven fiscal years beginning with 1947–48 as follows: 12.5, 28, 18, 0, 14, 12, and 12 percent, respectively.[7]

The Political Geography of Hong Kong

A singular combination of circumstances in Hong Kong has brought about prosperity without much internal pressure for the welfare state or much outside intervention, which might have closed down this haven of economic freedom.[8] How has this come about?

Hong Kong offers more than most residents in the colony could expect or receive elsewhere. Refugees from mainland China have found physical security and an opportunity for economic improvement. They do not clamor for more state intervention in their personal lives;

many of them fled from an oppressive Communist government to obtain the freedom found in Hong Kong's noninterventionist society. The middle class has, on the whole, enjoyed prosperity in a booming private sector. Local and overseas investors continue to benefit from the low taxes and economic freedom that cannot be found in other Asian countries.

In turn, a prosperous free-trading and financial center populated largely by persons of Chinese origin, who constitute a ready-made export market for the Chinese mainland, has served the economic interest of the Communist government in Peking. Finally, entrepreneurs from the United Kingdom and other countries benefit from commerce in Hong Kong. Let us examine, in turn, this tripod of consent to a haven of economic freedom: Britain, China, and the local people.

The Mother Country. Hong Kong became a British possession in 1841 for the purpose of trade with China; it prospered as a free-port trading post specifically because it offered security and freedom from interference with trade. The early inhabitants, especially the Chinese, had little or no interest in politics; the colonial government, mainly concerned with providing a framework within which trade could flourish, did little more than maintain law and order and raise taxes to pay for the cost of the civil establishment and necessary public works.

Since the end of World War II, Her Majesty's Government has willingly, but on occasion reluctantly, guided her colonial territories through a process of self government within the Commonwealth to ultimate independence. Local pressures on the British to hand over independence, a key element in this process, have been missing throughout the entire post-war period in Hong Kong. It is a fair estimate of public opinion that the overwhelming proportion of Hong Kong residents clearly prefers British colonial rule to that of China. Independence is not a feasible option because of China's control over the political status of Hong Kong. The arguments in support of colonial government combine past and present strategic considerations, moral issues, and economic advantages.

Hong Kong was occupied in 1841 both for its commercial importance and as one of a string of British naval stations around the world that provided bunkering and repair facilities. This strategic era has now largely passed and the naval dockyard has closed down. There is no British fleet in the Far East and the British base in Hong Kong is now an isolated outpost, which as recently as 1974 underwent a manpower reduction. These forces remain not so much to defend Hong Kong against external aggression as to assist in internal security.

What is Britain's moral responsibility to the people of Hong Kong? With some exceptions, it has been British policy neither to hold back any territory that wished to become independent nor to push any territory faster than it wished to go. But Hong Kong has no referendum or electoral machinery by which Her Majesty's Government can assess the wishes of the majority of the people. Still, I do not know a single knowledgeable analyst of Hong Kong who argues that a majority want an end to colonial government.

Finally, the economic rationale for British colonial presence. Hong Kong is not a captive market for British goods; almost all British goods have to compete in Hong Kong's free market with those of other countries. A number of benefits flow to Britain in the form of pensions paid to retired Hong Kong civil servants in Britain, dividends paid to British shareholders in Hong Kong firms, and payments for commercial facilities arranged through the City of London; but the amount of these benefits does not on its own make colonial rule profitable to Britain. Finally, as a dependent territory of the Crown, Hong Kong was formerly required to keep its official government reserves, and the greater part of the reserves of the banking system, in the form of sterling—which constituted an external source of support for the pound in Britain's overall trading and balance-of-payments accounts. This practice terminated in 1974 when the British government unilaterally ended its guarantee of the value of Hong Kong's sterling assets (at a rate of US$2.40 to one pound). The Hong Kong government and the banking system are now free to invest their official reserves and assets into any currency, and they do so on the basis of comparative rates of interest they can earn elsewhere and the comparative security of other currencies. Thus Hong Kong reserves, which have at times been equivalent to between one-quarter and one-third of Britain's total gold and foreign exchange reserves, no longer constitute a benefit for the mother country. The one residual advantage is that the nationalized British Airways Corporation gains from Britain's authority to negotiate landing rights at Hong Kong's Kai Tak airport. This is so because Britain manages the colony's external affairs, and therefore landing rights at Kai Tak are typically granted to foreign carriers only if British Airways is given preferential foreign routes. The colonial status and geographical location of Hong Kong give British Airways a strong negotiating position on profitable international air routes.

Although these historical, strategic, moral, and economic considerations apply with limited force in the late 1970s, still Britain has no compelling reason to withdraw, since all segments of the local population favor the stability that colonial government provides in Hong Kong. And, it is the colonial status that allows the Hong Kong government to follow what I regard as responsible fiscal and economic policies.

The Chinese Connection. The history of Hong Kong is in some ways a chronicle of rising and falling trends of trade and population in response to events taking place in China. Hong Kong's rapid population increase between 1945 and 1956, for instance is largely attributable to a massive exodus from China of refugees who sought freedom and asylum in Hong Kong. Similarly, industrialization is attributable to the inflow of a large labor force and capital from China during the turbulence of 1949 and 1950.

Most of Hong Kong's territory is leased from China. The New Territories were signed over to Britain in 1898 on a ninety-nine-year lease. Earlier treaties had ceded Hong Kong Island, Kowloon, and Stonecutter's Island to Britain in perpetuity, but these territories were not politically or economically viable as a separate unit. Without water and industry, which are located mainly in the New Territories, the foundation of Hong Kong's economic prosperity would disappear.

The post-1949 Communist government in Peking has renounced the "unequal treaties" that constitute Britain's claim to sovereignty in Hong Kong, clearly stating for the United Nations Special Committee on Colonialism and Decolonization that the settlement of Hong Kong's political status is an internal Chinese matter; it does not fall within the category of colonial territories. China thus denounces both the New Territories lease and the earlier cessions of territory as invalid, yet simultaneously acknowledges colonial authority in Hong Kong.

Until China chooses to renegotiate with Britain, Hong Kong provides economic benefits to China. China seeks foreign exchange. Since the mid-1960s, receipts from Hong Kong are believed to have accounted for some 40 percent of China's total earnings of foreign exchange.

China supplies Hong Kong with about 20 percent of its imports, including a wide range of inexpensive consumer goods, oil products, the bulk of its food imports, and a substantial quantity of water. It buys little in return, leaving a huge balance-of-payments surplus which underwrites China's development policies. In 1974, for example, Hong Kong imported HK$5,991 million worth of goods and services from China and exported, in return, only HK$296 million, leaving a favorable balance of HK$5,695 million (about US$1.2 billion). In addition, Hong Kong is the clearing house for remittances to China. Local and overseas companies and individuals remit to their relatives and business associates in China more than US$50 million a year.

Another important aspect of Chinese-Hong Kong trade is the colony's function as a redistribution center for Chinese-made goods. Hong Kong has the largest, deepest, and most modern port facilities along China's coast. Several hundred million dollars' worth of Chinese goods are annually exported through the port of Hong Kong for worldwide destinations.

Quite apart from these quantifiable economic benefits are the indirect, but very tangible, benefits that Hong Kong provides as a convenient center for trade contacts, financial negotiations, and access to Western technology.

China's special relationship is thus predominantly economic. Ideological consistency dictates that Hong Kong, as a dependent territory of a sovereign foreign power on Chinese soil, should not exist at all. But the Chinese will allow the colony of Hong Kong to continue so long as it serves the wider interests of the Chinese regime. One cannot forecast, though, how and if a change in that profitable relationship with Hong Kong will alter future appraisal of continued colonial status in Hong Kong.

The Local People. The colony's government operates under two very specific constraints. One is the need for the continued support of British public and parliamentary opinion, and the other is the continued confidence of both local and overseas businessmen and bankers. Both props might collapse in the face of a restless and discontented population.

The fact is that the vast majority of the people of Hong Kong are content with their lot. The last twenty years have witnessed only three serious outbreaks of rioting. The first riot in 1956 was a faction fight between Kuomintang and Communist supporters over the flying of Nationalist flags on October 10th, the Republic of China's national day. The second riot of April 1966 erupted from the Star Ferry's proposal to raise its first class fare five cents (one American penny); this was the only riot specifically directed at government policy. Finally, the disturbances of 1967 were a direct spillover of the excesses of the Great Proletarian Cultural Revolution across the border, and ended once stable conditions returned to China.

Urban workers show little inclination to protest or even organize in legally permissible ways. Unions contain fewer than 14 percent of the economically active population, and strikes are infrequent. Disinterest in union activity mirrors the broader pattern of political apathy. Approximately four hundred thousand persons were eligible to register for the voters' list in the March 1973 Urban Council elections—Hong Kong's only broadly elected body—but only 31,384 did so. Of these, only 8,765 bothered to vote—28 percent of the registered electorate and barely 2 percent of the eligible electorate. The upper strata of the

Chinese population typically channel their activism into philanthropic or community associations, or seek to be nominated for public service on the colony's appointed councils rather than into groups with a specifically political orientation.

Why is this colonial government so readily accepted, albeit apathetically, given the evidence of postwar political history? Note first that a large proportion of Hong Kong residents, some 42 percent, are not Hong Kong born, but are instead refugees from the various provinces of China. Most refugees, seeking comfort and security, are predisposed to political quietism; they have little interest in any form of political agitation. Living in Hong Kong makes one present-oriented: few can insure how long Hong Kong will survive, and all know that the wishes of the local inhabitants will not figure prominently in Peking's deliberations.

Too, life in Hong Kong has been materially good. Between 1964 and 1973 the index of real wages increased 58 percent, and this prosperity attenuates potential discontent over colonial rule. I was repeatedly told by many Chinese students at the University of Hong Kong that the purpose of Hong Kong is to make money. Hong Kong has no other public, moral, intellectual, artistic, cultural, or ethical purpose as a society of individuals. It is just one big bazaar. It is not uncommon for articulate student radicals to forget their radical aspirations shortly after finding well-paid jobs on graduation. The only reward for successful political activity is an unpaid seat on the relatively unimportant Urban Council.

Finally, it is fair to characterize the performance of the Hong Kong government as exceptionally efficient. Nearly half the population live in subsidized housing at below market rents, and great strides have been made in the development of medical and health services: the infant 'ity rate fell from 99.6 per 1,000 live births in 1950 in 1974. The last case of smallpox was reported over twenty years ago, and few cases of cholera have been reported for the past seven years. Roads, communications, port facilities, water supply, and public utilities have also kept pace, all financed by revenue yields from a very low tax rate.

Since most Hong Kong residents presume that Hong Kong's future will be determined by ministers in Peking and London, there is little scope or point to local political activism. Rapid economic growth is what life in Hong Kong is really all about. Individual opportunity for gain provided in Hong Kong's free market economy allows the local people to subscribe to the Jeffersonian maxim of "that government is best which governs least." The well-to-do malcontents can emigrate overseas, and those less fortunate can freely cross the border back into China. Many well-educated persons have resettled abroad, but virtually no one has crossed the border at Lo Wu. However, thousands each year flee China for Hong Kong, either legally or clandestinely. The barbed wire is on the Chinese side of the border, not the Hong Kong side.

The Open Economy

Let us briefly return to economic realities. Recall that Hong Kong's economy is open at both ends. Approximately 90 percent of Hong Kong's manufactured goods are exported, and the overwhelming bulk of its foodstuffs and consumer goods are imported. It has little control over restrictions or quotas that other nations may implement against its goods, nor over embargoes that may restrict both importing and exporting. This open structure restricts the range of possible government economic intervention: the government can do little to alter the cost/price structure of exports or imports to the benefit of Hong Kong. The fact of Hong Kong's externally dependent economy, due to its almost complete resourcelessness, encourages a hands-off attitude by government toward the private sector. I've delighted in meeting a good many British expatriates who have come to appreciate the "automatic corrective mechanism" after just a few short months in Hong Kong. The typical expatriate normally laments on arrival that the government lacks compassion in its laissez faire attitude towards the inegalitarian outcomes of the market. Few are as enthusiastic as new converts. These same expatriates now warn against the folly of walking in Britain's economic footsteps, and even suggest that the proper remedy for the poor health of Britain's economy is to turn over the administration of the United Kingdom to the Hong Kong government.

Summary

The social, economic, and political context of Hong Kong's market economy has comprised this first discussion. We have examined the political geography of Hong Kong and have found it to rest on a tripod of consents: British royal authority, Peking's recognition (dare we say approval?) of colonial government, and the tacit consent of the Hong Kong people. These are the cornerstones of political and economic stability.

Political stability is not by itself a sufficient condition of a market economy. The fact of Hong Kong's heavy dependence on external trade reduces considerably the scope for government's internal management of the economy. In this context, it is both reasonable and efficient to insist that external price movements be freely transmitted to the internal cost/price structure. But political stability and external dependence do not jointly insure that a government will practice both non-interventionism and responsible fiscal and monetary policies. For this we must add yet another dimension.

The political geography of Hong Kong and the realities of an externally dependent economy are two of the incentives and constraints under which the Hong Kong government determines its economic and fiscal policies. Next we look at the peculiarities of Hong Kong's anomalous practice of public finance and non-interventionism to see why its officials spend and tax less than their counterparts in other industrial societies.

Politics and Economic Freedom

II.

In spending only what it can afford, the Hong Kong government is, by worldwide standards, unique. Except for a small number of tropical paradise tax havens, no other government so intently holds expenditure within means. Its standard rate of tax on earnings and profits is the lowest in the industrial world, and its official government reserves are the largest in proportion to any year's expenditure. It can do this because of its colonial status, which frees public officials from the demands and interests of politically active citizens and candidates—demands that often stretch public expenditure beyond public means.

The colony's present-day public finances stand in sharp contrast to countries in which overspending, deficit financing, extensive use of loan finance, governmental control over the economy, and political instability are the rule. Circumstances in Hong Kong reveal a different landscape of financial and economic life:

1. The practice and policy of balanced budgets; 2. The maintenance of fiscal reserves, which have historically ranged from one-half to one year's expenditure; 3. The almost complete avoidance of public debt; 4. Habitual underspending by government departments; 5. A widespread economy ethic throughout government; 6. A general aversion to central

planning; 7. The commercial provision of public economic services; and 8. Minimum intervention in, or regulation of, the private sector.[1]

I want to explore the incentives, constraints, and ideological commitments that encourage Hong Kong's unique fiscal and economic responsibility. We shall examine, in turn, the historical precedents of financial administration in Hong Kong, the constitutional underpinnings of its practice, and the practice and philosophy of balanced budgets and nonintervention. Nor do I want to leave out the personalities and competence of Hong Kong's financial administrators. Hong Kong may be among the world's most highly personalized governments. The reason is that all lawful power lies with the Governor: none of his staff has specific constitutional power or authority, apart from the Judiciary. Personality attributes thus affect the balance of actual authority between high-ranking subordinates as they run the administrative machinery of government. We shall, therefore, stress the position and personality of Hong Kong's Financial Secretaries, the men who dominate the economic dimension of public life.

The Beginnings of Economic Freedom

Although Britain first occupied Hong Kong in January 1841 in the wake of the Opium War with China, a colonial administration was not officially set up until the Nanking Treaty was ratified on 26 June 1843.[2] Very soon thereafter, the colony began to play an influential role in the China trade and attracted a substantial multiracial community of foreign traders. But the opium trade also gave rise to a community of unruly residents and to crime. The first urgent problem of the Hong Kong government, then, was to maintain law and order and protect persons and property.

At the outset, Hong Kong was a barren island with no large or settled community entitled to political representation. It was established as a military, diplomatic, and trading station, not as a settlement in the normal sense. Because of this, the Secretary of State for War and the Colonies imposed firm imperial control on the new colony; self-government was not and never has been in Hong Kong's political cards.

Administrative absolutism meant, in practice, that the colonial government did little more than maintain law and order—and raise taxes to pay for the cost of the civil establishment and necessary public works. Each of these tasks was accomplished with some difficulty. An early proposal to tax imported wines and spirits met with unanimous opposition of the local Legislative Council, whose members said that such a tax would be contrary to Hong Kong's free-port doctrine. Indeed, merchant opposition at that time set a precedent which applies with equal force more than one hundred years later whenever new or increased taxes are proposed. Local Hong Kong merchants repeatedly complained of the revenue measures of successive Governors, condemned as unconstitutional an ordinance to impose rates on property, and called for the home government to subsidize colonial expenses since Hong Kong was held for imperial interests connected with the whole of the China trade.

Britain was not responsive to merchant protests. Parliament instructed Hong Kong's Governors to defray from colonial resources all public expenditures save the salaries of three principal officers in the colonial government. In 1855 Sir John Bowring, the Governor, happily announced that Hong Kong had reached the goal of complete self-support. Although on occasion future Governors sought assistance from the home government, the doctrine and practice of self-support was entrenched in Hong Kong financial procedure.

Colonial Regulations

Hong Kong, as a Crown Colony, was administered under the colonial regulations.[3] The regulations date back to 1837 and serve as "directions to Governors for general guidance given by the Crown through the Secretary of State for the Colonies," especially in financial and administrative matters. Governors and their officials were not considered very free agents of the Secretary of State, who was empowered to fill all important vacancies in the colonial services, maintain disciplinary control over these appointments, and impose standards for strict financial discipline. In particular, the annual estimates (the budget) of each colony required the Secretary of State's approval well before the beginning of the financial year. The colonial regulations specified expenditure of approved funds, clarified how and when variations in the approved estimates could be allowed, and instructed local officials on the submission and auditing of public accounts. It was the responsibility of the Secretary of State and his aides in the Colonial Office to see that colonies did not run into debt and impose a charge on the British Treasury.

Colonial Office policy and the colonial regulations that applied to Hong Kong visibly reflect the prevalent economic theories of nineteenth-century Britain, which stressed the passive role of government in the economy. Private individuals and companies, not the government, were held responsible for the creation of wealth. Although the British Treasury ceased its oversight of colonial expenditures after 1870, Colonial Office approval was typically required until the 1930s, when convention began to replace the letter of the regulations and the estimates were no longer submitted to the Secretary of State before presentation to the local legislature. Of course, the Secretary could still use his powers of disallowance if he thought that a colony's spending plans would cause recourse to the United Kingdom Exchequer.

Until financial autonomy was granted to the Hong Kong government in 1958, the 1951 colonial regulations provided the formal background against which the financial authorities in Hong Kong had to work. The guiding principle was that all colonies should aim to be self-supporting and contribute to the cost of their own defense. This lengthy set of regulations, specifically those given over to questions of finance in Chapter VIII, the longest chapter, insured financial transparency. These regulations resemble the financial and accounting regulations under which the present Hong Kong government works.

The principles and practice of public finance in present-day Hong Kong are, in the main, a direct offshoot of colonial tradition. Hong Kong continues to follow the pattern of colonial accounting practices because they provide a good set of rules for financial integrity. Where they have been refined, it has nearly always been in the direction of increased clarity. The form and scope of the budget have changed little since the granting of financial autonomy in 1958, and still resemble the traditional colonial regulations. In addition, nineteenth-century values of economic liberalism still influence official thinking and practice in Hong Kong.

Government business in Hong Kong is structured according to the Governor's *Regulations of the Hong Kong Government*,[4] and he has the authority over their interpretation and application. The chapter on financial regulations reflects the substance and flavor of the parent colonial regulations. The commitment to fiscal integrity is neatly illustrated in regulation 308, paragraph (1): "It is the responsibility of Heads of Departments to exercise strict economy in the votes under their control. Money must not be spent, if it is not necessary to spend it, simply because it has been voted."

Constitutional and Administrative Framework

To begin with, Hong Kong is not a representative democracy. Administrative and executive authority lies in the hands of appointed civil servants whose staff members, at the higher levels, have been recruited chiefly from the United Kingdom.[5] Only one of the top ten positions in the Hong Kong government is held by a locally born Chinese; only a handful at department head rank or higher (some forty positions) are locally recruited persons. Neither periodic elections nor public opinion polls guide or constrain the administrative decisions of these appointed officials. Hong Kong political activity is not party competition, the quest for votes, popularity, high standing in the polls, a share of the "pork barrel," or public debate of social issues. It is rather day-to-day decision making by appointed officials, sometimes within the administration, often in consultation with one of a myriad of official advisory committees, or, on occasion, with the open soliciting of the public's views.

Constitutional authority for making policy is concentrated in the Governor, assisted, in practice, by his Executive Council. The Governor's powers are defined by the *Letters Patent and Royal Instructions to the Governor of Hong Kong*.[6] He is, as representative of the Queen, the head of government, and constitutionally accepts his instructions from the Secretary of State, though, in practice, instructions are rarely given.

As chief executive, the Governor has the final responsibility for the administration of the colony. He makes laws by and with the advice and consent of the Legislative Council. In the execution of the Governor's power and authorities, the *Royal Instructions* stipulates that he shall consult with the Executive Council—his advisory body consisting of both *ex officio* members of government and other official and unofficial persons appointed by the Secretary of State on the Governor's recommendation. However, the Governor is the sole constitutional authority for administration, legislation, and finance. He is specifically empowered to act in opposition to advice given him by Members of the Executive Council. No provision exists for formal voting in the advisory body; the Governor seeks to distill a consensus from the advice he is given by his Councillors and acts on this advice unless he has overwhelming reasons for not doing so.

The *Letters Patent and Royal Instructions* thus empowers the Governor with broad authority. His chief constraints are the reality of Hong Kong's political economy and his responsibility to Her Majesty through the Secretary of State, who, in turn, is responsible to the United Kingdom Parliament for the administration of colonies.

The *Letters Patent* also sets forth the constitutional foundations of the Governor's legislative authority. Clause VII stipulates that "The Governor, by and with the advice and consent of the Legislative Council, may make laws for the peace, order, and good government of the Colony." The Legislative Council consists of the Governor, its President, a number of ex officio and other government members, as well as a majority of nominated unofficial councillors, who are drawn from among the prominent members of the community.

Decisions of the legislature are typically consensual, with an occasional holdout or two among the appointed unofficial members. The Legislative Council rarely shows an inclination to withhold consent from legislation proposed by the official bureaucracy. Proceedings in the legislature rely heavily upon British parliamentary procedure in which the government proposes and the legislature disposes. The norm has been that official motions are unanimously accepted with little comment, though official and unofficial members often use the forum to speak on topics of personal interest.

There exists but one standing committee of the Legislative Council, the Finance Committee, which must approve all proposals involving the expenditure of public funds before presentation to the full meeting of the Legislative Council. Membership consists of three ex officio members, who do not vote, and all unofficial members of the Legislative Council. The standard view of this committee, and its two Public Works and Establishment Subcommittees, is that its members work hard, long, unpaid hours, but typically defer to the government's expenditure proposals.

Before moving into the bureaucratic corridors of Central Government Offices on Lower Albert Road—the headquarters of the colonial administration—it might be useful to summarize the colony's constitutional status. All power is concentrated in the Governor. None of his staff has specific constitutional power or authority. He is instructed, under the *Royal Instructions*, to consult his Executive Council in the exercise of his power and authorities, but may lawfully act in opposition to their advice. His legislative authority, under the *Letters Patent*, entails majority votes in the legislative chamber; however, he has the power to disallow proposed laws. In practice, he is never forced to do this. Serious differences of opinion are invariably reconciled before an official government measure is proposed to the Legislative Council. Remember, too, that all unofficial members of both councils are appointed from among prominent members of the community: none of these officials are accountable to an electorate, nor does the job pay a salary from official expenses. Thus, the economic and budgetary policies of the Hong Kong government are determined largely by the Governor and his high-ranking subordinates.

Bureaucratic Administration

Hong Kong's modern Western government—modern in the technical sense of bureaucratic administration, not in the sense of representative democracy—can perhaps best be classified as a "no-party administrative state" governed at the highest levels by civil servants appointed from overseas.[7] Many of these expatriate civil servants are headquartered in the Colonial Secretariat; their leader, the Colonial Secretary, serves as a chief of staff to the Governor and, under his direction, carries on the general administration of the government. However, the Secretariat has recently been reorganized along the lines drawn up by McKinsey and Company as presented in their May 1973 report, designed to restructure power within the Secretariat and promote overall efficiency in government.[8] The word "colony" has almost become taboo in official parlance: the Colonial Secretary is now the Chief Secretary and the Colonial Secretariat is typically Central Government Offices. The Queen, however, is still the Queen. And I shall use the term Secretariat as a shorthand for Central Government Offices.

The Secretariat.[9] Overall, the Secretariat is responsible for the coordination of the whole bureaucratic structure. Responsibility for the workings of the Secretariat, along with each branch and department of the public service, resides in the Chief Secretary, the designated head of the Civil

Service. Within the Secretariat is also one other officer of similar rank, the Financial Secretary. Although he ranks second in seniority to the Chief Secretary, he is, within his own sphere, responsible directly to the Governor. These two secretaries hold the highest positions in government and may often advise the Governor on policy matters.

The Secretariat is divided into three main parts: (1) "policy" branches, (2) financial and economic, and (3) establishment. The responsibility of the "policy" branches is to screen departmental policy proposals, coordinate departmental policies, facilitate the work of departments, and contribute, where possible, to policy formulation in program areas under their overall supervision. The financial and economic division is directed by the Financial Secretary. The Economic Branch supervises those government departments which provide economic services directly to the public, and also oversees the whole field of financial and economic policy. Finance Branch devotes much of its efforts to the preparation of the annual budget. Finally, the Establishment Branch runs and staffs the Civil Service for about forty government departments.

The Finance Branch. Many officials in the Hong Kong government, when referring to the fifth floor of the Secretariat, use the phrase "the all-powerful Finance Branch." Day-to-day administration is directed by the Deputy Financial Secretary, who coordinates preparation of the annual budget, authorizes increases in departmental budgets, and sits as a member on a host of committees concerned with finance. Finance Branch is a large and well-structured branch of the Secretariat and dwarfs in size, scope, and importance the "policy" branches.

Finance Branch has three distinct capabilities: (1) advice to departments and policy branch heads on the feasibility and financial implications of proposed policies, (2) preparation of papers to seek the Legislative Council's Finance Committee approval in light of these financial implications, and (3) preparation, in conjunction with the departments, of the annual estimates of expenditure and revenues to carry out the government's official policies for the forthcoming fiscal year. All finance in Hong Kong is *centralized*: although Finance Branch consults with the departments before preparing papers for Finance Committee approval, the substance of the papers may not necessarily have the department head's full concurrence. Finance Branch is especially watchful that no financial request constitutes a new policy of government. Regulations of the Hong Kong government require that all requested increases or decreases in departmental expenditure be fully justified, and if the increase or decrease is based upon a shift in policy or the implementation of a new policy, justification entails prior policy approval from the Governor and his Executive Councillors or, for minor matters, Secretariat approval.

Finance Branch and its leader, the Financial Secretary, have played an increasingly crucial role in government since about 1935, when the colony's financial and economic affairs began growing in complexity and importance. Finance Branch personnel generally speak with one voice, subscribe to a uniform value-for-money ethic of financial administration, and enjoy a constitutional position of maintaining central financial control over public spending. As one spokesman in the Finance Branch made clear to me:

> "In this economy [Hong Kong] you cannot divorce the financial implications of any policy from the policy itself, and the Financial Secretary must be in on

> the determination of all policies. No Financial Secretary could do his job in Hong Kong under any other circumstance."

The Financial Secretary. In addition to his responsibility for preparing the annual budget estimates, the Financial Secretary and his aides are heavily involved in wide-ranging activities that bear upon the fiscal health of government and the economic health of the colony. Any government policy with fiscal or economic implications must involve the Financial Secretary. Given that Hong Kong's economy is open at both ends, with the great bulk of its manufactured goods exported and its consumer goods imported, any policies that tinker with the fiscal or monetary system of Hong Kong could have dangerous economic repercussions. The Financial Secretary must, in Hong Kong's economic circumstances, be a key policy maker; he is, by definition, the master of the budgetary process.

What kind of men have sat in the Financial Secretary's seat? A cross-section of Hong Kong's intellectual, administrative, and economic elite, some fifty-one persons I interviewed in the fall of 1973 as part of my research into policy making and public spending in government, uniformly praised the abilities of these men of finance, particularly Sir John Cowperthwaite [Financial Secretary, 1961–1971] and Philip Haddon-Cave [Financial Secretary, 1971–present]. One official noted that the constitutional basis of centralized finance, conjoined with Cowperthwaite's intellectual skills, afforded his arguments clear dominance within the Secretariat and the policy making councils.

> "Cowperthwaite was brilliant, well-trained in economics, suffered no fools, and was highly principled. He wouldn't last five minutes in a similar post in Britain, since he was not predisposed to compromise

any of his principles—only the constitutional structure of Hong Kong allowed him that power."

Another said: "The relative strength of different departments in the Secretariat is a matter of personalities. And Cowperthwaite was both strong and bright." And an even more emphatic evaluation by a Chinese junior administrative officer left no doubts: "Cowperthwaite was brilliant." From the department perspective, "Cowperthwaite controlled the purse strings and decided priorities." Yet it was precisely this all-powerful central financial control that underpinned one officer's exclamation: "We owe our prosperity to one man: Cowperthwaite." A majority of these same respondents now assert that Haddon-Cave is the most competent and brilliant civil servant in Hong Kong since Cowperthwaite's retirement. Without question, the Financial Secretary is the lynchpin in the exercise of preparing the colony's annual budget.

Hong Kong's people have been blessed by the competent performance of their Financial Secretaries. These men are responsible in law to the Governor of Hong Kong, and in conscience to their principles and the economic welfare of the people of Hong Kong. Let us now look at the economic and budgetary policies of the Hong Kong government through the principles of economic and financial administration enunciated by successive Financial Secretaries.

Economic and Budgetary Policy

Historical precedent has shaped Hong Kong's economic and budgetary policies.[10] The incentives and constraints that dictate contemporary practice derive, in the case of budgetary policy, from the letter and spirit of the colonial regulations—the emphasis on self-support and balanced budgets. In the case of economic policy, the free port status of early Hong Kong discouraged an interventionist government: wealth creation was to reside in the private sector. Nonintervention and fiscal responsibility are still the cornerstones of a modern administration which, in the presence of an otherwise increasingly restrictionist world beset with deficit budgets and growing public debts, must surely seem anachronistic.

Economic Policy

In Hong Kong, economic affairs are conducted in an environment of virtually unfettered free enterprise. Government policy has long dictated a virtually hands-off approach toward the private sector, an approach that seems well suited to Hong Kong's exposed and dependent economic and political situation. The philosophy that underlies government in Hong Kong can be summed up in a few short phrases: law and order, minimum interference in private affairs, and the creation of an environment conducive to profitable investment. Regulatory economic controls are held to a minimum, no restrictions are placed on the movement of capital, little protection and few subsidies are given to industry, and the few direct services provided by government are operated on a commercial basis.

Capital Movements. Hong Kong is a completely free market in money. No barriers restrict exchange between the Hong Kong dollar and other currencies. Hong Kong is thus a genuine financial haven: financial assets are easily transferable or convertible with minimum regulation or interference. Indeed, the ever-increasing funds that have been attracted to Hong Kong banks helped finance industrial development. These capital inflows have been so substantial throughout the postwar period that Hong Kong is today a major financial center. Hong Kong's Financial

Secretaries have uniformly believed that guaranteeing the free flow of capital is the only way to insure its continual movement to Hong Kong. Before the first reading of the 1955–56 appropriation bill, Arthur Clarke, then Financial Secretary, posed the following question: "Can we as a Government do anything to maintain or to improve this economy of ours?" His reply:

> "We can do little to encourage capital to come here or to remain here, apart from two things; firstly, to place as few restrictions as possible on its movement; and secondly, to allow it to have as a good a return as possible by keeping the rate of tax on profits low."[11]

Sir John Cowperthwaite echoed the same belief in the 1963 budget debate. He sought to allay fears that the absence of restrictions on capital movements might lead to a harmful outflow of funds into overseas securities:

> "My honourable Friend F. S. Li has further suggested that something might be done to prevent the outflow of funds into investment in foreign securities. I myself do not think it would be wise to try to do anything about this. We enjoy a considerable net inflow of capital and I am sure that a condition of its coming, and staying, is that it is free to flow out again. It is also important for Hong Kong's status as a financial centre that there should be a maximum freedom of capital movement both in and out. I am sure that we make a substantial net gain."[12]

Subsidies. Except for land grants in the mid-1970s to land-intensive industries that inject new technology into the economy, and the segregation of industrial land to protect it from the competition of other land users, no protection or government assistance is traditionally given to manufacturing industries, utilities, service industries, or private citizens. No attempt is made to distort factor prices in favor of any particular type of development. Market forces are allowed to shape the economy, and industries that lobby for protection from the competitive forces in the marketplace are fiercely resisted.

Postwar Financial Secretaries have each resolutely opposed any system of subsidy that exempted any industry from paying the full costs of the resources it consumed. Each has also opposed subsidies for the well-to-do. In 1960, for example, Arthur Clarke rejected a proposal to rebate water charges for industry. Private motorists have been subjected to a continuous series of price increases for parking meters and public garage charges to recover rising costs in full. Housing subsidies for all but the very lowest income families have been rejected on grounds that building housing for lower-middle and middle income groups diverts resources from a maximum housing effort for those at the lower end of the income scale. Cowperthwaite also points to the absurdity of housing subsidies for middle-income families:

> "One of my difficulties about differential public housing standards arises from the fact that all public housing is subsidized by concessionary grants of land, by low interest rates and by long capital redemption periods; and in consequence the more expensive the accommodation that is built or the lower the densities of occupation, the higher, other things being equal, the subsidy to each individual; that is, *the better off a family is the higher the subsidy it tends to receive; which is surely absurd.*"[13] (italics mine)

Government Economic Services. What part does government play in the operation of the Hong Kong economy? Its philosophy is consistently noninterventionist, and its reliance on the private sector and the market mechanism for the production of national wealth extends even to public utilities and public transport: electricity, gas, the telephone services, buses, ferries, and the tramways are all entrusted to private hands, though they derive monopoly franchise rights under an ordinance that requires the provision of adequate services and governmental approval of fare increases. Government, however, intervenes in such areas as the provision of water, land ownership and public housing, and operation of the airport and the railway, and regulates, in part, banking, rents for domestic premises, pollution emissions, labor conditions, the rice trade, fish marketing, and so forth.

Government's management style is governed by the philosophy of self-support except, as in the case of such specific subsidies as housing and education and income assistance to individuals, where it has overriding social reasons not to do so. These social exceptions aside, government seeks to operate its economic services on a commercial basis. Once it has determined that it must supply a service to achieve social or economic objectives, either because the provision of services cannot be found in the private market or because these are common facilities that only the public sector can provide (e.g., water supplies), it tries to conduct its affairs with minimum cost to the general taxpayer.

The waterworks, managed entirely by government, have always posed the problem of self-support and the self-liquidation of capital investment at commercial rates. The historical difficulties with provision of water—there are no rivers and only a few large streams in the colony—have led to an annual concern with excess demand, rationing, and correct pricing. The colony's water consumption is supplied by collecting and storing rainwater through systems of catchwaters and reservoirs, a desalinization scheme, and the purchase of water from China. The uneven incidence of rainfall, combined with a lack of space for storage, has often resulted in shortage, leading on occasion to water rationing.

The 1963 rationing ordeal brought on discussion of an adequate supply of water in the 1964 budget debate. Cowperthwaite's analysis is an excellent exposition of price theory, opportunity cost, and the law of demand.

> "This leads me to another water topic—the appropriate daily period of supply we should be aiming at. I find myself considered inhumane or unprogressive or sometimes merely odd, by some of my colleagues as well as members of the public, when I suggest that it is not axiomatic that a twenty-four-hour supply in all circumstances must be our immediate aim. I cannot myself see any grounds for the belief that a twenty-four-hour domestic water supply is an inalienable right of civilized man. It may be, *if he can afford it and is prepared to pay the price*. But here we come up against the problem of determining the public's preferences and priorities in the spending of its money and also of determining the true costs to the community. I would not take issue with a doctrine that a twenty-four-hour supply should be available if charged for at commercial costs, because they measure the true cost to the community in terms of choice between competing uses of scarce resources. And at commercial prices a twenty-four hour supply may in fact be quite easy to provide because on a commercial basis rates would be very high, probably several times those calculated on the present noncommercial basis, and consumption would therefore be substantially

> lower. But clearly charges at this level would be a hardship for those who are contented with less and lower prices; and I think there must be, in the case of a public utility like water, an overriding principle that the rate should be low enough for everyone to pay for enough to meet his reasonable needs. What *cannot* in my view be justified is investment of very large amounts of public capital (for which there are so many competing demands) for the provision of a supply for maximum consumption on the basis of normal Government pricing at noncommercial interest rates and redemption periods.[14]

Cowperthwaite's predecessor, Arthur Clarke, raised water rates in 1960 from 80 cents per thousand gallons to one dollar for the purpose of "inducing those concerned to be more careful in their use of water—to use less water."[15] Cowperthwaite raised water charges in 1965 and again in 1971, when he rejected the principle of subsidizing free provision of water to all users:

> "I doubt if it were ever proper to raise taxes for the provision of water free, or on a subsidized basis, to all users. I see no reason, for example, why someone who is content with an economical cold shower should subsidize someone who is able to luxuriate in a deep hot bath; or why someone who waters a few plants in pots on his windowsill should subsidize someone who waters his extensive lawns; or why the careful domestic user should subsidize the profitable operations of a large industry. It is for these equitable reasons, as well as *the general principle of charging for a commodity what it costs*, that suggestions that taxes . . . should replace water charges must be deemed inappropriate.[16] (italics mine)

In his 1974 budget, Haddon-Cave also stressed the importance of recovering costs from government economic services. He noted that:

> "...we must vigorously update, and keep up to date, our system of fees and charges for those goods and services which are not a fair charge to general tax revenue; and this means ensuring that the full cost of these goods and services is covered.
>
> ...we cannot afford to allow social and political consideration unduly to influence our management of public utility undertakings, namely, the waterworks, the post office, the airport, and the Kowloon-Canton Railway.
>
> ...our system of licenses must be updated, and kept up-to-date, not so much on a historical cost basis, but in terms of what the market will bear."[17]

And, on the persistent question of water rates, he warned:

> "The consequence of all this is that the cost of water looks as if it is going to be appreciably higher than my predecessor anticipated when he determined the present charges. And as I said earlier, we simply cannot afford to subsidize our utilities from General Revenue."[18]

Government's commercial management style for the waterworks also applies to other economic services. The budgets are full of explanations for the application of commercial practices to the provision of car parks, the railway, the airport, post office, and so forth.

"One trouble is that when Government gets into a business it tends to make it uneconomic for anyone else."[19] These Cowperthwaitian comments are the very heart and soul of the government's economic policy: hands off the private sector, to avoid the risk of discouraging enterprise, and provision of government economic services, when required, on a commercial basis.

> "For I still believe that, in the long run, the aggregate of the decisions of individual businessmen, exercising individual judgment in a free economy, even if often mistaken, is likely to do less harm than the centralized decisions of a Government; and certainly the harm is likely to be counteracted faster. . . .
>
> . . . It has to be recognized, and it is recognized over a large part of our daily life, that the community's scarce economic resources *can be efficiently allocated only by the price mechanism.*[20] (italics mine)

Budgetary Policy

The principle of self-support and balanced budgets in the colonial regulations applies equally to a financially independent Hong Kong. The government follows a fiscal policy intended to avoid sustained or systematic deficits and, if possible, accumulate reserves that permit sustained government expenditure over a long recession without serious cutbacks, and, in the process, earn interest to meet recurrent expenses. After noting a small deficit in the immediate postwar budget, Table 4 shows that the budget has ended the year in surplus in twenty-eight of the succeeding thirty-one years through 1978. Moreover, these surpluses are computed after charging against current revenue all capital expenditure other than a comparatively small amount financed by borrowing; government has traditionally financed its entire public works program and other capital investments from recurrent revenue. Annual capital expenditure has approached HK$2 billion (US$400 million) in the mid-1970s.

Table 4: Revenue and Expenditure of the Hong Kong Government, 1946–1978 (in HK$ Millions)

Financial Year	*Revenue*	*Expenditure*	*Surplus* (+) *Deficit* (−)
1946–47	82.1	85.6	−3.5
1947–48	164.3	127.7	+36.6
1948–49	194.9	160.0	+35.0
1949–50	264.3	182.1	+82.1
1950–51	291.7	251.7	+40.0
1951–52	308.6	275.9	+32.7
1952–53	384.6	311.7	+72.8
1953–54	396.9	355.4	+41.5
1954–55	434.5	373.3	+61.1
1955–56	454.7	402.5	+52.3
1956–57	509.7	469.5	+40.1
1957–58	584.2	532.7	+51.5
1958–59	629.3	590.0	+39.4
1959–60	664.6	710.0	−45.3
1960–61	859.2	845.3	+13.9
1961–62	1,030.4	953.2	+77.2
1962–63	1,253.1	1,113.3	+139.8
1963–64	1,393.9	1,295.4	+98.5
1964–65	1,518.3	1,440.5	+77.8
1965–66	1,631.7	1,769.1	−137.4
1966–67	1,817.8	1,806.1	+11.7
1967–68	1,899.5	1,766.0	+133.5
1968–69	2,081.1	1,873.0	+208.1
1969–70	2,480.7	2,032.2	+448.5
1970–71	3,070.9	2,452.2	+618.7
1971–72	3,541.0	2,901.0	+640.0
1972–73	4,936.0	4,300.0	+636.0
1973–74	5,240.8	5,169.2	+71.6
1974–75	5,875.3	6,255.2	−379.9
1975–76	6,519.5	6,032.2	+487.3
1976–77	7,493.0	6,591.0	+902.0
1977–78 (estimated)	9,235.0	8,160.0	+1,075.0

Source: *Hong Kong Statistics 1947–1967*, Table 9.11; *Supporting Financial Statements and Statistical Appendices for the Estimates of Revenue and Draft Estimates of Expenditure 1977–78*, p. 832; and *The 1978–79 Budget*, paragraph 63.

Budgetary policy in Hong Kong is unique in industrial states. Where else can one find low taxes, budget surpluses, substantial government reserves, a virtual absence of public debt, and a vast increase in public services?

A first step is to understand the Financial Secretary's attitude toward the budgetary process. Witness, for example, Haddon-Cave's 1972 statement on the relation between fiscal policy and economic circumstances:

> "Sir, I am sure honourable Members will have noticed that both Sir John Cowperthwaite and Mr. Clarke always dealt with the revenue estimates before the expenditure estimates and for obvious reasons: in our circumstances, there are severe limits to the range of indirect taxes which can be imposed (e.g., a customs tariff, such an important revenue raiser elsewhere, would be quite inappropriate in our circumstances); and there are severe limits also to the marginal rate of direct taxation. By and large, therefore, we must fit public expenditure to available public resources and not extend those resources to fit expenditure.[21]

The avoidance of systematic or continuous deficit financing has been pursued with almost religious fervor in order to avoid Keynesian-type inflationary finance. The two serious annual deficits in postwar Hong Kong, which occurred during the 1965–66 and 1974–75 fiscal years, helped contribute to an increase in the rate of tax on earnings and profits in 1966 from 12.5 to 15 percent and a small 10 percent surcharge on profits tax to begin for 1975–76 tax liabilities. These were deemed necessary and timely to support an accelerating rate of expenditure. Cowperthwaite emphatically rejected deficit financing in his 1963 budget:

> "But I will *not* be proposing a course which has been under public discussion recently—deficit financing. It

is wholly inappropriate to our economic situation. In its least extreme form it is based on the theory that additional money generated by a Government deficit (and given currency, as necessary, by use of the printing press) will stimulate consumption and thereby production, in time to match the excess money with goods before real inflationary harm is done. Unfortunately we don't, and can't, produce more than a small fraction of what we consume, and increased consumption would merely mean increased imports without matching exports; and a severe balance of payment crisis, which would destroy Hong Kong's credit and confidence in the Hong Kong dollar; and which we could not cure without coming close to ruining ourselves. Keynes was not writing with our situation in mind. In this hard world we have to earn before we spend."[22]

Government Reserves. The Hong Kong government is surely unique among industrial states in that it possesses very large reserves, either deposited with the local commercial banking system or held overseas, which have historically ranged from one-half to one year's recurrent and capital expenditure, although this fraction has declined in the last few years to range from 34 to 44 percent. Existence of these large reserves is, in and of itself, anomalous in a world of chronic public debt and deficit financing. How has this reservoir of official funds arisen in a postwar society which saw extensive spending on public works and social services? Few governments ever attain the goal of a balanced budget, much less the accummulation of large surpluses.

In Hong Kong, several factors are at work: a recognition that long-term economic forecasting is unreliable locally, conservative estimating of revenue, dramatic increases in revenue from rapid economic growth, and a never-ending concern over increasing recurrent commitments.

A balanced budget requires an accurate prediction of both estimates of revenue and expenditure. Revenue estimates are especially troublesome, as they vary directly with the performance of the economy. Hong Kong's exposed situation frustrates, almost precludes, reasonably accurate forecasts of economic output and, hence, revenues; thus the natural tendency is caution. Moreover, given general recognition of Hong Kong's unpredictable economy, heads of departments might reasonably be expected to prepare and submit extremely conservative estimates of revenue. Caution is preferred to reckless abandon. Better to accumulate reserves than face possible cutbacks in overexpanded and inflated government programs predicted on a greater revenue basis.

Taxation. The last element in the panorama of budgetary policy is taxation. Hong Kong has, to my knowledge, the lowest standard rate of tax on earnings and profits of any industrial state. On this point, as on other aspects of economic and budgetary policy, Hong Kong's financial czars preach with consistency. This result is paradoxical, but comforting. The official line is Gladstone reincarnated: a narrow tax base and low standard rates of direct taxation facilitate rapid economic growth which generates high and ever-increasing tax yields. These revenues, in turn, finance an extremely ambitious program of public expenditure on housing, education, health, and welfare services, and on other forms of social and community services, with virtually no need to resort to loan finance. Each Financial Secretary repeats the message: Investment is stimulated by low rates of direct taxation, limiting charges on profits and earnings to activities within Hong Kong, and imposing no tariffs or controls on the movement of capital.

Not only do low taxes stimulate investment and rapid economic growth, in the official view, but also the absence of a full income tax and its associated inquisitorial powers is deemed equally virtuous. (Details of taxation are set forth in the next chapter.)

Monetary System

Prior to 1935, Hong Kong, along with China, was on a silver standard. When China abandoned the silver standard in 1935, Hong Kong followed suit and adopted a pound sterling exchange standard. In 1972, Hong Kong again changed standards, this time switching from pounds sterling to U.S. dollars. On 26 November 1974, Hong Kong abandoned the last of these pegged exchange rates and shifted to a floating exchange rate.

Why have these changes occurred? The answer to this question can once again be found in relation to the colony's trading arrangements and its economic, fiscal and monetary policies. We have already seen the cardinal goal of economic policy to be that of generating maximum output from the colony's resources of land, labor, and capital, and that efficient use of these resources demands freedom of trade in an externally dependent economy. Public spending is dictated by the self-imposed constraint of a balanced budget, weighted on the side of fiscal conservatism and the accumulation of fiscal surpluses. Finally, the goal of monetary policy is to inflate the domestic currency as little as possible to minimize distortions to the local cost/price structure. Before 26 November 1974, it was possible to realize all three of these goals under a fixed exchange rate regime. With the advent of massive world inflation, especially in Britain and the United States, it is no longer possible to maintain a fixed exchange rate regime and avoid domestically generated inflation through increases in the domestic money supply.

The details of Hong Kong's monetary history tell a unique and interesting story. To begin with, private banks print money under their own name. With the exception of a small but growing issue of coins (including $1,000 gold coins) by the government, Hong Kong's currency notes are issued by three commercial banks, the Hongkong and Shanghai Banking Corporation, the Chartered Bank, and the Mercantile Bank Limited, though from 1978 on the latter will no longer issue notes. Hong Kong does not have a central bank.

Hong Kong's money supply process is presently on a *fiat standard* in which money consists of pieces of paper and rights to pieces of paper. The note-issuing banks will, in exchange for a $10 bill, issue a new $10 bill, two $5 bills, or some token coins, but do not offer any physical commodity. This money is valuable to a person because other people will accept it in exchange for valuable goods, or currencies of other nations, and not because of any intrinsic value in and of itself.

The value of this currency is regulated by an Exchange Fund, which was set up by the government in 1935 when the Hong Kong dollar ceased to be based on silver. This fund, held by government, in effect creates money when it receives payment from the note-issuing banks in exchange for certificates of indebtedness denominated in Hong Kong dollars which provide the legal backing for the notes issued by the banks. These certificates are non-interest bearing and are issued and redeemed at the discretion of the Financial Secretary. The Exchange Fund deposits with the banks the notes received for the certificates of indebtedness, which are then treated as liquid assets of the banking system, on which banks can create multiple credit expansion.

The Exchange Fund then deploys its resources in a variety of investments, both long- and short-term, denominated in several currencies, and can use these assets to intervene in the foreign exchange market.

Formerly, the Hong Kong money supply process rested on a commodity standard, that of silver. On a commodity standard, money consists of a physical commodity, such as gold or silver, or rights to a certain number of ounces of that commodity. Prior to 1935, Hong Kong currency consisted of Mexican silver dollars, which were declared to be the colony's legal tender in 1842, British trade dollars minted in India from 1895 (equivalent to the Mexican dollar); and locally issued bank notes (backed almost completely by silver, bullion, or sterling securities held in bank vaults or with the Crown Agents) that increasingly became by 1890 the customary means of payments, although they were not legal tender. It was simply too inconvenient to deal with large amounts of silver.

When Hong Kong abandoned the silver standard in 1935, the ordinance that established the Exchange Fund required the note-issuing banks to surrender all silver previously held by them against their note issues in exchange for certificates of indebtedness. The exchange value of the Hong Kong dollar was then set at 1*s* 3*d* sterling (or HK$16 to one pound). Under this pound sterling exchange standard, the relationship of the Hong Kong dollar with sterling was maintained by the operations of the Exchange Fund in conjunction with the three note-issuing banks. The three banks could increase their issue of currency notes when they purchased certificates of indebtedness from the Exchange Fund. The banks paid 1*s* 3*d* to the fund for the right to issue a Hong Kong dollar note. In turn, the fund stood ready to buy back certificates of indebtedness from the three banks at a rate of 1*s* 2 7/8*d*. So long as the Financial Secretary did not call in outstanding certificates, the system was largely automatic. When the banks believed that the demand for currency was rising, they could purchase with sterling the necessary certificates and issue more banknotes. Conversely, if the three banks felt their cash holdings were excessive, they could return some certificates to the Exchange Fund for sterling. The Exchange Fund kept its assets in pounds sterling and thus the domestic currency issue was fully backed by external assets held in the Exchange Fund. Since the Exchange Fund earned interest on its sterling assets, the domestic issue enjoyed a minimum external backing of no less than 105 percent under this arrangement.

The exchange value of the Hong Kong dollar with sterling was not statutory, but came to be generally regarded as fixed. As a dependent territory of the British Crown and a member of the sterling exchange control area, Hong Kong was required in practice to keep its money reserves, including the greater part of the reserves of the banking system, in the form of pounds sterling securities.

The transition from a *commodity* standard based on silver, to a pound sterling exchange standard gave way, in turn, to a U.S. dollar exchange standard. The link with sterling weakened when the pound was devalued by 14.3 percent against the U.S. dollar in November 1967. At that time the value of the Hong Kong dollar was set to a rate of HK$14.55 to one pound—a corresponding devaluation of only 5.7 percent relative to the U.S. dollar. When the pound was allowed to float downwards in June 1972 and the sterling area largely disbanded, the Hong Kong government switched from sterling to U.S. dollars to fix the exchange value of the Hong Kong dollar. When the U.S. dollar itself was devalued by 10 percent in February 1973, the rate of HK$5.65 to US$1 was adjusted to a new rate of HK$5.085 to US$1. On 26 November 1974, the Hong Kong government no longer used the assets of the Exchange Fund to support the Hong Kong dollar at the official, pegged rate and let it float upwards from this official rate under pressure of heavy selling of U.S. dollars. Its exchange value has ranged as high as HK$4.60 to US$1 in recent times. Hong Kong has thus made the transition from a monetary system pegged to a fixed exchange rate to one of floating rates.

What forced a shift to floating rates in Hong Kong? A cardinal goal of monetary policy is to avoid internally generated inflation and, as a corollary, to insulate the local economy from external inflation. So, the reason that a community like Hong Kong must shift from fixed to floating exchange rates is to sever its money supply from those of other nations that suffer over-rapid monetary expansion with consequent high rates of inflation. In previous years, Hong Kong could maintain a balanced government budget, expedite foreign trade in a most efficient manner, and inflate its currency as little as possible. This is no longer possible when two of its major trading partners, the United States and Britain, suffer rapid monetary expansion and concomitant inflation.

Under the fixed exchange rate system in effect before November 1974, the domestic money supply was largely determined by the balance of payments. Under a fixed rate, the banks and the government (through the Exchange Fund) accept, passively, the net outcome of all overseas transactions; if Hong Kong earns more foreign exchange than it uses, then one or both sectors purchase the excess and thereby increase the Hong Kong dollar deposits held by the private sector. Thus under a fixed exchange rate system, any surplus in foreign exchange is typically converted into Hong Kong dollars and this increase in the domestic money supply generates internal inflationary pressures.

When the exchange rate floats, as it has since late 1974, the acquisition of foreign assets by the bank and the Exchange Fund is a matter of *choice* by both private and public sectors, not by the automatic passive acceptance of an excess or deficit of foreign exchange at a fixed rate of conversion. After the banks and the government have each adjusted their own holdings of foreign assets, any difference between the supply of and demand for Hong Kong dollars is met by an adjustment in the exchange rate, or the price, of the Hong Kong dollar.

Thus when the exchange rate floats, the domestic money supply is not so automatically determined. Under floating rates, the government commands greater discretion in the determination of the money supply than under the former system of pegged rates. It has chosen to adopt this policy of discretionary intervention, in lieu of the former system of a virtually automatically determined money supply based on the balance of payments, to avoid the importation of overseas inflations with its damaging distortions in the local cost/price structure. It has chosen to let the Hong Kong price level adjust to world prices through the mechanism of a floating exchange rate rather than through an increase in the domestic money supply, with the consequent effect of an increase in local prices.

Role of Public Policy

It would be wrong to say that government adheres dogmatically to a laissez–faire style of management in all of its public economic activities. The state is the ground landlord in Hong Kong and spends between 15 and 20 percent of the national income providing roads, compulsory primary education, extensive medical and health services, subventions for numerous social welfare agencies, and public housing for about 45 percent of the population, a proportion that is scheduled to increase steadily over the next decade.

A first and very important activity that government must undertake derives from its constitutional authority to maintain law and order, carry on overseas commercial relations, and dispose of assets vested in the Crown. Let us take the disposition of Crown assets to illustrate how government tempers even its constitutional authority via the market mechanism.

In a series of letters and instruments, all dated 4 January 1843, addressed by Lord Aberdeen to Sir Henry Pottinger, Hong Kong's first governor, a number of temporary measures for the administration of Hong Kong were set forth. The governor was to make grants of land

> "[except] that H.M. Government would not choose that . . . grants should be made to parties whose object in obtaining such grants would be to dispose of them again with advantage to themselves. . . . He further told that if, as a result of the establishment of a free port and the introduction of those liberal arrangements by which foreigners would be encouraged to come, a great commercial entrepôt were created, then H.M. Government would feel justified in securing to the Crown the increased values that the land would then have.[23]

All land in Hong Kong is thus owned by the Crown. Except in the New Territories, which the British still acknowledge as encumbered with a lease expiration date of 1997, all Crown leases are granted for seventy-five years, usually renewable for a further seventy-five years at a reassessed Crown rent. Basic policy, except the new industrial land policy for land-intensive industry, is to sell leases to the highest bidder at public auction. Land sales have ranged from a low figure of around HK$10 million annually before 1954 to a high of several hundred million during years of building boom (although revenue from land sales comprises a much smaller proportion of total revenue than in previous years). Industrial land is scarce in Hong Kong and must be rationed efficiently to ensure that each square foot generates the maximum income for the colony.

A second activity that government undertakes is the provision of services not found readily in the private market, for example, subsidized housing for low-income families. In 1953, in order to cope with the housing shortage created by an influx of refugees from China, the government began a crash building program. For squatters who illegally occupied land required for development, accommodations known as resettlement estates were built. Many persons housed in resettlement estates had been compulsorily evicted from squatter housing. As of 1976, about 44 percent of the population lived on public housing estates and the Housing Authority was slated to build housing for another 1.5 million people by 1984, at a cost of HK$6,000 million at 1976 prices.

Government housing officials argue that the private sector has been unable or unwilling to provide adequate low-rent housing since competing forms of investment, especially luxury flats, normally offer higher rates of return to capital. Government provision of low-rent housing, with rents far below open market rentals, does not encourage private supply of mass market housing, thus, in turn, leaving government with the self-imposed responsibility of housing higher and higher percentages of the population at moderate rents.

A new and growing activity of government, and the one on which we conclude this discussion of departures from a laissez-faire mold of administration, is provision of social security. The chief feature of social security is the Public Assistance Scheme, which is non-contributory and provides a guaranteed income to needy households, along with a rental allowance for costs of accommodation. A Disability and Infirmity Allowance Scheme provides a cash allowance on top of any public assistance payment, to encourage the severely disabled and the elderly infirm to remain within the community as long as they are able to do so. These two scales of payment are indexed to keep pace with the cost of living.

Summary

I have titled this discussion "Politics and Economic Freedom," but it is apparent that the correct title is "The Absence of Politics and Economic Freedom." Historical application of the colonial regulations and the free port status of early Hong Kong supplied the initial constraints on budgetary and economic policy. The Colonial Office imposed the principle of balanced budgets and self-support; the absence of tariffs and other restrictive measures offered little opportunity for constructive intervention by the colonial government in Hong Kong's economy. Moreover, British economic policy in the nineteenth century strongly reflected the influence of Adam Smith's invisible hand: both wealth creation and distribution were to reside in the private sector.

The constitutional status of contemporary Hong Kong allows public officials to respond to internal and external economic forces, rather than to special-interest political groups. We have designated Hong Kong "a no-party administrative state," that is, a modern polity without elections and parties. Public officials in Hong Kong do not respond to the incentive of periodic popular elections, a necessary short-run view of political decision making.

Rather, career civil servants, free from popular short-run pressures, can adopt a long-term view of economic efficiency. All power is concentrated in the position of the Governor, and economic and budgetary decision making is, by and large, delegated to his chief assistant, the Financial Secretary and his staff.

What, then, are the incentives and constraints that confront Hong Kong's financial authorities? They are, first, the reality of Hong Kong's externally dependent economy, in which the consequences of an incorrect decision by government quickly come home to roost. The fact of almost complete resourcelessness means that there is little the government can do to alter the internal cost/price structure of Hong Kong goods to the benefit of Hong Kong; the converse is not true. The government can, through intervention, alter the internal cost/price structure to the detriment of Hong Kong. Tariffs will raise internal costs of production and lower external competitiveness. High taxes will reduce Hong Kong's attractiveness as a center for investment. Subsidies will keep marginally inefficient firms in business, to the detriment of total national production. Noninterventionism as the cornerstone of economic policy overlays the ideology of the free-market economy with the reality of resourcelessness, thus encouraging a succession of Financial Secretaries to speak in the language of an eighteenth century Adam Smith and a nineteenth century William E. Gladstone.

The fact of resourcelessness also means that growth in the economy must be export-led. A growing public sector comes only at the expense of reduced private investment and income. For this reason, public spending has ranged from 14 to 20 percent of gross domestic product, and the financial authorities watch that public spending does not grow at a faster rate than the overall economy. Hong Kong has no plans for a central bank, and its international and domestic borrowings are restricted exclusively to self-liquidating projects at commercial rates of interest.

I should mention, also, that the political uncertainty over Hong Kong's future means that investments are typically amortized in very short periods: who will tie up money for twenty years in Hong Kong's exposed political geography?

I've had occasion to talk with the key administrative personnel in the Hong Kong government, especially its past and present Financial Secretaries. These are men who take enormous satisfaction in the economic well-being of the people of Hong Kong, despite vocal internal criticism and external British expatriate charges of inhumane and noncompassionate government. Yet their stubborn, but reasoned, adherence to policies of noninterventionism and fiscal integrity has served the people of Hong Kong well. Given the chance, these policies will do equally well in the future.

Doing Business in Hong Kong

III.

The view of Hong Kong I have thus far set forth is that of one political scientist, although it is tainted by my favorable endorsement of a colonial government's economic responsibility and financial integrity.[1] I see Hong Kong as a virtuous polity for three reasons: first, political geography and economic realities dictate a hands-off policy toward Hong Kong's market economy—there is little the government can do to alter the cost/price structure of imports or exports to the benefit of Hong Kong; second, the constitutional status of the colony, expecially the centralized position of finance, allows its decision makers to take a long-term view of the economy—they are free from the ever-present electoral pressures that prevail in economic decision making in most democratic polities; and, third, the ideology of Hong Kong's key administrators supports a market bias and a belief in the efficient properties of the pricing mechanism in allocating resources. This three-dimensional image of public virtue explains why Hong Kong's people live and work in the industrial world's freest economy.

I'd like, now, to remove my academic garb and don the attire of a businessman. I want to look at Hong Kong from the standpoint of those who work, invest, and make money in Hong Kong. This perspective reveals why both local and overseas investors find Hong Kong so attractive. Our review of Hong Kong's economic magnetism will range across topics such as its convenient geographical location, general business requirements, taxation, employment and labor unions, manufacturing conditions, banking and finance institutions, transportation and port facilities, and some personal reflections which have emerged over six trips to Hong Kong in the last fourteen years.

Location

A glance at a map of the Asia-Pacific area shows how convenient Hong Kong is as a center for many activities within the region. It is a focal point for all major air and sea routes. It is the preferred base for the large majority of companies with regional headquarters in Asia. It operates the largest container terminal in the Asia-Pacific area. Finally, more than one million tourists visit Hong Kong each year. From the businessman's standpoint, Hong Kong is an ideal location for the China trade, both as point of entry for the biannual Chinese Trade Fair held in Canton and for the conduct of business with Hong Kong branches of various Chinese state trading corporations. Hong Kong has been and remains a gateway to Southern China, astride major air and sea routes of the Far East. These advantages of location provide enormous economic benefits that more than compensate for Hong Kong's uncertain tomorrow. Remember, the British Crown Colony of Hong Kong flourishes on what is admittedly Chinese soil and granite, to be restored, someday, to its rightful owner.

General Business Requirements

The government of Hong Kong does not impede the setting up of private business enterprise.[2] Government rigorously enforces the principle of free entry into almost every line of production. Legal formalities required to set up a business, even to establish a private company are, for the most part, few and inexpensive. The Business Registration Ordinance of 1952 reduced the annual fee for a business license from HK$200 to HK$25, exempting certain small or dormant businesses from the fee altogether. Recently, the annual fee has been raised to HK$150, but select small businesses are still exempted. For approximately thirty American dollars, one can lawfully open a business in Hong Kong.

Incorporation is also inexpensive: the fee is HK$370 plus $4 per $1,000 of authorized capital. Foreign corporations with branches in Hong Kong pay neither incorporation nor annual registration fees; their obligations are simply to file with the Registrar of Companies a copy of the company's charter, its by-laws, its annual financial statement, and the names of persons in Hong Kong who are authorized to receive notices served on the company.

Finally, anyone who opens a factory or industrial undertaking that employs more than twenty persons must, prior to registration, comply with Fire Service Department safety standards and Labor Department standards for workers' safety. Approval is, in normal circumstances, routinely quick.

It is a general principle of Hong Kong economic and tax policy not to discriminate between residents and non-residents. On this principle, overseas investors may fully own local factories. At the end of 1975, overseas interests either fully or partly owned 271 factories. Overseas capital employed a total labor force of 59,607, or 8.7 percent of all workers in Hong Kong's manufacturing industries; this capital represents a direct investment of about HK$1,730 million (US$375 million). United States firms have the largest single bloc of foreign investment, which accounts for about 14 percent of Hong Kong's exports. A number of statutory and private organizations actively encourage new or expanded overseas investment, and the Immigration Department cooperates in producing an attractive business climate: although overseas persons require an employment visa to work in Hong Kong, foreign employees of multinational or international companies are routinely accepted.

Taxation

"No taxation without representation" is a famous American battle cry. King George III had imposed a threepence-on-the-pound tax on tea, a tax equal to 1.25 percent of its value. Over this threepenny bit, the colonists fought a revolutionary war. "No taxation without representation" is not the war chant of Hong Kong's 4.5 million people; "no taxation" will do just fine, thank you![3]

The tax collector in Hong Kong is not an inquisitor. The Inland Revenue Ordinance does not levy an overall income tax; instead, it levies four separate direct taxes on profits, salaries, property, and interest and *restricts taxes to income arising in or derived from Hong Kong*. Hong Kong has negotiated no tax treaties with other countries, including the United Kingdom. The colony grants no tax incentives for overseas investors, nor makes any distinction between residents and non-residents. Thus, a resident is exempt from taxation on profits received from abroad while a non-resident must pay tax on profits arising in Hong Kong.

The maximum tax rate on profits is 17 percent, after allowing as deductions all expenses incurred in producing chargeable profits. I think it correct to say that Inland Revenue liberally interprets allowable deductions. Historically, the standard tax on profits stood at a flat rate of 10 percent in 1947, rose to 12.5 percent in 1951, and then to 15 percent in 1966. A 10 percent surcharge added in 1975–76 remains in force.

Salaries tax is also limited to income arising in or derived from Hong Kong. The tax is computed on a progressive basis, ranging from a minimum of five percent on the first assessable $10,000 to 30 percent on taxable income exceeding $50,000, after deducting personal allowances. However, the total tax cannot exceed the standard rate of 15 percent of total income.

There is no system of pay-as-you-earn withholding for salaries tax. The Government makes available Tax Reserve Certificates which are redeemable at 4.2 percent tax-free interest, if used to pay taxes. Or, private arrangements can be made with any bank to withhold a certain sum every month for tax purposes. I suppose many live at the margin each month and simply take their chances at the race track when taxes are due. The point is that it is the individual's responsibility to make his own arrangements to pay his required salaries tax, and this approach to the taxpayer's responsibility is in keeping with the general themes of individual initiative and individual responsibility so prevalent throughout Hong Kong.

Property tax is a charge on the owner of property, payable by the person who pays the rates under the Rating Ordinance. The Commissioner of Rating and Valuation sets the tax at 15 percent of rateable value, after a 20 percent deduction to allow for repairs and maintenance. Owner-occupants are exempt from the charge, though each is limited to only one primary residence.

Interest tax, unlike other taxes, is collected at source; the payer of the interest deducts the tax and hands it directly to Inland Revenue. The tax is uniformly applied at the standard rate of 15 percent, and is limited to interest which arises in, or is derived from, Hong Kong.

An interesting feature of the Inland Revenue Ordinance is the right of a taxpayer to opt for what is known as Personal Assessment, under which he aggregates his income from all sources but gets the benefit of the same personal allowances and sliding scale of tax as is allowed for salaries tax purposes. Personal assessment thus reduces interest and property tax liabilities of low-salaried taxpayers.

Some other minor taxes raise, in all, several hundred million dollars. Estate duty is imposed on that part of a deceased person's estate which is situated in Hong Kong. The rate of duty varies from six percent on estates valued at $400,000 to 18 percent on those in excess of $3 million.[4]

Among the few remaining taxes is stamp duty, which imposes fixed duties on certain classes of documents; its yield fluctuates with turnover on the stock and property markets. Entertainments tax is levied on admission charges to movie theatres and race meetings, the former at 10 percent of admission price and the latter at 22 percent. Betting duty is imposed on bets made on an authorized totalisator or pari-mutuels and on contributions toward authorized cash sweeps. The duty is recovered from the Royal Hong Kong Jockey Club, which holds the monopoly for conducting such operations, including a limited form of off-course betting. Finally, a Hotel accommodation tax is levied at the rate of four percent on hotel and guest house charges.

Overall, two-thirds of government revenue comes from direct and indirect taxes, with one-third from fees and charges, a ratio that has remained constant for nearly twenty-five years. Within the tax system, earnings and profits tax yields have grown as a proportion of total revenue. In 1949–50, for example, the ratio of direct to indirect taxation was 23:77; by 1974–75, the ratio had shifted to 56:44. This chronological decline in the share of indirect taxes reflects the determination not to impose customs tariffs or excise duties, which might seriously distort production costs and thereby weaken Hong Kong's external competitiveness. Too, levies on consumer goods and foodstuffs might increase government revenues, but would simultaneously increase the cost of living.

Yields from earnings and profits taxes are income-sensitive and rise most rapidly when Hong Kong enjoys sustained periods of vigorous growth, as in the postwar period; conversely, yields from indirect taxes are much less income-sensitive.

It is interesting to review the history of direct taxation in Hong Kong.[5] An income tax was first proposed in 1939, but it was widely opposed and the compromise outcome was the appointment of a committee, known as the War Revenue Committee, which recommended in 1940 a partial income tax based on certain sections of the 1932 Ceylon Income Tax Ordinance. However, the committee did not favor a permanently staffed tax collector's office; instead, they called for temporary staff to avoid any burden on the colony's pension commitment. In reservation, two members of the committee warned that the Ordinance might reduce capital flows into Hong Kong. They moreover stressed that a proposed Business Profits Tax was inquisitorial in nature and should thus be omitted from the bill. In any case, the committee's recommendations constituted the 1940 War Revenue Ordinance and its postwar successor, the Inland Revenue Ordinance. A key feature in the 1940 Ordinance was that only profits arising in, or derived from, Hong Kong should be taxed.

To return to the past, the War Revenue Ordinance imposed taxes on property, salaries, and profits, but low yields saw a reconstituted War Revenue Committee unfold in 1941, chiefly to consider ways to combat apparently widespread evasion. To reduce opportunities for evasion, the committee considered, but did not recommend due to great opposition, a full income tax. Instead, they proposed a tax on interest, which has been the only major extension of the direct tax net in the past thirty-five years.

Need for additional postwar revenue to balance the colony's budget led to a minor recasting of the War Revenue Ordinance in the form of the Inland Revenue Bill. Two new concepts in the bill were the "standard rate" and personal assessment. A first Inland Revenue Ordinance Review Committee, appointed in 1954, made only minor technical recommendations, leaving the underlying principles of the Ordinance intact. A second review committee, appointed in 1966, offered several minor substantive and administrative suggestions and, in particular, tried to get a better handle on the problem of tax evasion. Incidentally, bribes paid to public officials can be entered as a legitimate business expense, itemized as "gifts" and "entertainment" in the audited accounts filed with the Inland Revenue Department. The Commissioner of Inland Revenue notes that under Section IV of the Ordinance, the police cannot obtain any information from him on the illegal payments.

Although minor changes have been made in recent years, two essential features of the Inland Revenue Ordinance persist: the territorial limitation of the charge and the non-aggregation of different forms of taxable income, except at a taxpayer's option. A third review committee, appointed in the 1975–76 fiscal year, was charged to consider the basic structure of the Inland Revenue Ordinance. It suggested compulsory aggregation of total income, but recognized that this recommendation would require lengthy evaluation.[6]

Employment and Labor Unions

Hong Kong's remarkable productivity rests, in part, on its hardworking and adaptable labor force.[7] That the Chinese worker is diligent goes without saying; reinforcing this diligence is the fact that rising real wages available in Hong Kong's market economy have markedly improved the ordinary worker's standard of living and obviated any need for trade unions, at least in the eyes of the local work force. Supply and demand for labor, not trade unions, determine wage rates.

Hong Kong labor is chiefly engaged in small enterprises. At the end of 1975, for example, only 40 factories employed over 1,000 workers; conversely, 28,000 factories had fewer than 49 workers, of which 20,000 engaged less than 10.

Hong Kong does not impose a statutory minimum wage. Earnings of industrial workers fluctuate with overall economic activity, although it is customary to award each worker an extra month's salary at Chinese New Year. Loyalties to firms are less important than salary and fringe benefits, and thus workers respond quickly and rationally to alternative market opportunities. In the face of an economic downturn, it is customary among workers to accept a reduction in hours rather than force lay-offs and "break the next man's rice bowl."

The intelligent application of labor legislation both encourages enterprise and allows wage rates to reflect market conditions rather than arbitrary bureaucratic decisions. In response to growing pressure from Her Majesty's Labor Party government, Hong Kong has enacted new and revised labor legislation. However, many of these ordinances reflect current market realities for sick leave, vacation time, and fringe benefits and should not be viewed—for the time being at least—as obstacles to productive enterprise.

Labor legislation consists chiefly of the Factories and Industrial Undertakings Ordinance, the Employment Ordinance, and the Workmen's Compensation Ordinance.

The first of these imposes no restrictions on the working hours of men in industry. Children and women, however, are restricted. Children under fourteen years of age are excluded from industrial work. Those aged fourteen to fifteen qualify for industrial work, but may not put in overtime. Finally, all women and children aged sixteen to seventeen may work no more than 200 hours of overtime per year.

The Employment Ordinance governs terms of employment for all full-time manual laborers and all non-manual workers earning less than HK$2,000 per month. It provides ten annual statutory holidays; stipulates sick leave, maternity leave, and monthly rest day conditions (although an employee may request to work on his four monthly rest days); states the frequency, forms, and place of wage payments; and, finally, guarantees the right of any employee to join a trade union.

Lastly, the Workmen's Compensation Ordinance provides personal compensation from injury or death arising from work.

Overall, the Hong Kong government is generally lenient in its view and enforcement of labor legislation, suspecting that British complaints of "sweated labor" reflect protectionist sentiments of British unions in conjunction with the textile and garment industries, rather than sincere humanitarianism. A growing economy, what life in Hong Kong is really all about, does more to improve wages and working conditions by increasing the demand for labor than legislative and bureaucratic regulations which, by raising the cost of labor,have precisely the opposite effect.

That the labor force is generally content to accept the verdict of market forces is seen in work interruption statistics. In 1975, for example, only seventeen work stoppages transpired with but 17,600 man-days of work lost from an annual total exceeding 180 million man-days (a fraction of

1/10,000th). Investors do not confront any serious problems of strikes, work stoppages, or worker grievances. They need only pay the market price for labor to hire the requisite staff.

Trade unions play little part in setting wages or working conditions. Indeed, it is curious that the government of Hong Kong seems most active in trying to encourage unionism. One old China-hand claims that local Chinese workers see Western unionism as an obscure alien trick of Western democratic processes.[8] In terms of Asian living standards, the average Hong Kong worker is better off than all but the Japanese. Time-worn complaints of "sweated labor" and enslavement of little children in hothouse factories, periodically trotted out by British critics, better reflect British politics and ideology than actual working conditions in Hong Kong.

Manufacturing

The counterpart of employment and trade unions is manufacturing.[9] Among the elements that go into manufacturing, land is perhaps the most scarce and highly prized commodity in Hong Kong. All land in Hong Kong belongs to the British Crown and freehold tenures, save one historical exception, are never granted. A lease is typically given for seventy-five years, renewable for a further seventy-five years, for land on Hong Kong Island and the urban areas of Kowloon, south of Boundary Street. In the New Territories and in Kowloon north of Boundary Street, leases are written to expire on 27 June 1997, three days before the termination of the original ninety-nine year lease that granted the New Territories to the Crown in 1898.

Government makes land available to private developers and industrialists through public auctions. Again, local and foreign residents are treated equally and each may hold title to the lease. Land goes to the highest bidder, plain and simple. Building covenants, attached to land sale conditions, require a specific minimum expenditure within a reasonable period of time. The covenant prevents any individual or company from acquiring enormous banks of undeveloped land.

One can purchase land either by a lump sum or through several installment plans with varying rates of interest. Investors can also obtain land in Hong Kong from present holders of titles to a Crown lease. Land prices in Hong Kong reflect prevailing economic conditions; real estate prices have fluctuated both up and down in recent years, although industrial areas have greater price stability than the wide, volatile rising and falling margins associated with residential or commercial land and premises.

Several new policies allow investors to acquire land, apart from public auctions or purchase from another Crown lease holder, for the purpose of establishing intensive and heavy industrial projects. One long-term policy is the development of industrial estates, to be managed by a Government-funded Industrial Estate Corporation. Prospective occupants include industries that produce industrial raw materials, manufacture heavy machinery, and make appliances in the medium-heavy engineering industry. Premises will be sold on a square-foot cost basis or rented on five-year leases.

The second departure from public auctions is the Modified Land Policy which presumably accommodates industries that are unable to operate within industrial estates or in high-rise factories. Land is awarded to companies, by private treaty or by tender, if the applicant (1) is new to Hong Kong or upgrades technically an existing industrial process, (2) provides high-skilled employment opportunity for male workers, (3) is land-intensive and unsuitable for multi-story industrial buildings, and (4) supplies basic raw materials to other manufacturing industries in Hong Kong. Two large sites on Tsing Yi Island have already been sold under the Modified Land Policy to American firms—Dow Chemical Pacific and Outboard Marine International.

Trade restrictions against Hong Kong are of greatest concern to manufacturers in the garments and textiles industry. The flow of Hong Kong-made merchandise is controlled largely through foreign governments' imposition of quantitative restrictions and import duties. Hong Kong is a member of the General Agreement on Tariffs and Trade (GATT) and no longer accords preferential treatment to British and Commonwealth products. Both local and overseas exporters and importers receive equal treatment.

The most comprehensive restrictions confronting Hong Kong are the quantitative limitations imposed largely on its garments and textiles, the Multi-Fibre Agreement, which was created under the auspices of GATT. The great bulk of Hong Kong's garment and textile exports to European Economic Community countries and the United States is under export quota restrictions. These work as follows: importing countries specify quantities of each category of textiles and garments annually acceptable, with permissible growth rates in these quotas of about six percent a year. The quotas are then administered within Hong Kong by the Department of Commerce and Industry, which dispenses quota allocations on a formula basis.

Within Hong Kong itself, trade formalities are few and inexpensive. As a duty-free port, Hong Kong allows the entry and exit of most raw materials, consumer goods, and commodities with only a registration charge. To raise revenues, duties are imposed only on liquors, manufactured tobacco, and hydrocarbon oils. A handful of items require import and export licenses issued by the Department of Commerce and Industry for reasons of health, safety, or security. Certificates of origin are granted to manufacturers to qualify goods made in Hong Kong for entry under quotas, for Generalized Scheme of Preferences tariff rates, or for Commonwealth preferences. Cars are not subject to duties, but a first registration fee of 30 percent of their Hong Kong C.I.F. value must be paid to the Transport Department.

Finally, rent control protection, which is provided on some classes of residential accommodation, does not cover office accommodation.

Banking and Finance

The output of the banking and financial sector, a dynamic and growing element of the Hong Kong economy, is now equal to one-fifth of gross domestic product.[10] Because of political instability in Southeast Asia, the banking industry in Hong Kong has a somewhat different investment philosophy than do banks in Europe and America. Hong Kong banks make short to medium-term lending commitments, and have high degrees of liquidity. As previously noted, Hong Kong does not have a central bank, although the Hongkong and Shanghai Banking Corporation controls about 50 percent of the commercial banking sector and the concomitant domestic money supply.

Hong Kong's banking sector is regulated under the Banking Ordinance, which provides for a Commissioner of Banking who approves banking licenses and monitors the liquidity position of commercial banks. This ordinance was introduced in 1965 following a bank run in Hong Kong.

Banking licenses, the prerequisite to establishing a full-service commercial bank in Hong Kong, are issued only by the Commissioner of Banking. Since 1965, all new applications for bank licenses but one have been refused. The only exception that is made at present is to an application from a bank incorporated in one of Hong Kong's export markets and yet unrepresented by a bank in Hong Kong—subject, of course, to the criterion of reciprocity. It is also possible for foreign banks to acquire a major interest in a Hong Kong bank. As of March 1976, there were seventy-four full-service banks with 707 branches in business throughout Hong Kong; of these forty banks with 276 branches were incorporated abroad.

The Banking Ordinance also provides for representative offices of foreign banks in Hong Kong. In early 1976, eighty offices represented banks incorporated in many foreign countries. These representatives are not allowed to accept deposits or transact routine banking business, but are permitted to negotiate or arrange for transactions and loans on behalf of their head offices in their home country.

Hong Kong has recently experienced substantial growth in the operation of finance companies and merchant banks. These financial institutions offer many services normally rendered by commercial banks, but are specifically prohibited from opening checking accounts for their customers.

Finance companies concentrate largely on mortgage finance and installment credit for durable goods, and may not accept deposits from the general public for periods of less than ninety days. Merchant banks function as wholesale banking outlets for some foreign banks who do not enjoy local commercial banking licenses. Their most important activity is to arrange, or participate in, syndicated loans which finance government and private enterprises throughout the Asia-Pacific region. Merchant banks also offer foreign exchange service, lease financing, investment management services, and the underwriting of new share issues. Merchant banking business has boomed in recent years, commensurate with the growth in size and sophistication of Hong Kong's financial market and opportunities for project and development financing in the Asia-Pacific region.

Hong Kong is today the world's third largest gold market and operates totally free from government restrictions. Daily gold turnover on the Chinese Gold and Silver Exchange typically exceeds 350,000 ounces on an active day—more than the average daily volume on all U.S. gold markets.

Let me conclude this topic with a brief but, I hope, profitable dash through Hong Kong's stock markets. Hong Kong has no anti-trust laws, and the majority of listed companies on the several stock exchanges only issue about 25 percent of their paid-up capital to the public either by subscription or by private placement. Furthermore, Hong Kong does not require as full financial disclosure from public companies as regulatory authorities stipulate in either Britain or the United States. Both the *Far Eastern Economic Review* and the *South China Morning Post*, Hong Kong's leading newsweekly and morning English-language daily, often report stories of share manipulations or other unethical insider activities. These factors presumably contribute to the volatility of Hong Kong's stock markets.

One unique feature of Hong Kong's securities market is the settlement system. All security transactions must be settled in cash within the next working day. Any person wishing to sell short must be able to borrow securities for delivery on the following working day to complete the transaction.

Some Personal Observations

Borrowed Place–Borrowed Time by Richard Hughes tells us a lot about the pace of life in Hong Kong.[11]. Because investors are accustomed to getting a good return on their investment, things typically get done in a hurry. The idea that Hong Kong is a *borrowed place* living on *borrowed time* infuses all economic activity with an urgency of no tomorrow. Hong Kong is a place to make a living, but not a home.

In the December 1974 issue of *Passages*, contributor Tom Pickens titles his essay in the vivid language that accurately and forcefully describes the last great citadel of unrefined free enterprise. "Hong Kong is the only city in the world where you can have an idea at 9 o'clock in the morning, be in business at noon, and be making profits at 9 that night," Pickens tells his readers. He also says that imagination is the only limit to making money in this bustling community wherein business is transacted at a pace that makes New York look snailish by comparison. "In Hong Kong, laissez-faire is not just a quaint phrase from the past. It's a way of life."[12]

I submit that Hong Kong, among the world's more than 130 countries, most closely resembles the textbook model of a competitive market economy, encumbered only with the barest overlay of government. It is in this context that we should meet the pure form of *homo economicus*. His given name is *homo Hongkongus*.

Hong Kong man's first and most telling characteristic is his single-minded pursuit of making money. A companion characteristic is his emphasis on the material things in life. Hong Kong's free port, free trade economy offers for sale the latest in fashions, furnishings, food-stuffs, appliances, motor cars, gadgets, stereos—portable stereos, built-in stereos, automobile stereos, any and every conceivable brand and model of stereo at tax-free prices. If there is a new breakthrough in stereo goods to sell, some Hong Kong entrepreneur will be selling it that night. Tomorrow would mean foregone profits.

Material consumption and making money, or making

money and worldly goods is what life in Hong Kong is all about. The desire to acquire and accumulate as much money as possible in the shortest period of time: in Hong Kong, taxes do not discourage hard work. It so happens that I spent 29 August 1976, a pleasant enough Sunday, wandering about a very quiet London. What a sharp contrast it was to 5 December 1976, the Sunday I spent fighting the masses of Hong Kong for a space in the restaurants and stores. London was literally deserted and its citizens were clearly neither making nor spending money. Not so in Hong Kong. *Homo Hongkongus* works a twelve to sixteen hour day, seven days a week, almost 365 days a year—only Chinese New Year interrupts an otherwise single-minded obsession for making money. And Hong Kong's prosperous residents and overseas visitors have plenty of money to spend. I reckon there are more jewelry stores, good restaurants, and fashionable shops in Hong Kong's crowded streets than in any other city in the world—and why not, when the tax-free prices are taken into account. If economic man symbolizes competitive capitalism, he is alive, prosperous, and delightfully happy in Hong Kong. A quick glance across the border finds more than 800 million of his countrymen who ostensibly labor for love of ideology, not money or materialism.

Surely all is not well in Hong Kong. Would not a carefully designed survey of public opinion cast grave doubts on the soul of Hong Kong man. It would, of course, if we only interviewed the intellectuals, being absolutely careful not to talk with the ordinary working men and women. You see, apart from a few jobs in the universities and higher institutions of art and culture, the services of the literati are not in high demand. Indeed, the amount of high culture that Hong Kong's 4.5 million people demand is patently sub-optimal, that is, it is less than most Western middle class intellectuals, who are used to having their cultural tastes subsidized, would like. Higher education in Hong Kong is, for the most part, a path to self-improvement; if it civilizes in the process, well, O.K.

Hong Kong offers a limited cultural menu of art, music, and drama. Its critics call it a cultural desert and accuse it, correctly, of being an oversized bazaar. Hong Kong offers, in truth, exactly what the market will bear. When intellectuals complain about the lack of finer things in Hong Kong they are complaining, it seems to me, of three things: first, a limited demand for their services with corresponding low incomes; second, a failure of their fellow men (who are certainly less sophisticated and therefore need cultural tutelage) to share their tastes; and third, the government's unwillingness to subsidize their tastes at community expense, perhaps the gravest failing of the capitalist economy.

Dare I reveal my boorishness by saying that I find Hong Kong's economic hustle and bustle more interesting, entertaining, and liberating than its lack of high opera, music, and drama? East has indeed met West in the market economy. Chinese and Europeans in Hong Kong have no time for racial quarrels, which would only interfere with making money. The prospect of individual gain in the marketplace makes group activity for political gain unnecessary—the market economy is truly color-blind. Even harmony among ideological enemies lives in Hong Kong. The Hong Kong Hilton stands right across the street from the Bank of China.

There is simply no more exciting city on the face of the earth than Hong Kong—in large measure because it is the most robust bastion of free-wheeling capitalism. It may be, as is true of all human institutions, imperfect. I have previously criticized the Hong Kong government for its tendencies to increase spending on social programs, but can you name any other country which in 1976 enjoyed 18 percent real growth and only 3.4 percent inflation?

Is Hong Kong Unique? Its Future and Some General Observations about Economic Freedom

IV.

At midnight on Monday, 30 June 1997, the New Territories lease expires. At that exact moment, all territory north of Boundary Street, along with the several hundred offshore islands now administered by Her Majesty's colonial government, revert to China. Only Hong Kong Island, Stonecutter's Island, and Kowloon south of Boundary Street are ceded in perpetuity. Although the legal status of permanent cession is questionable, Hong Kong as we now know it is not economically viable without the New Territories, in which is located the bulk of its industry, water, and the airport, not to mention half its population. Hong Kong may or may not die on 1 July 1997, but the expiration date of the New Territories lease brings sharply into focus the key political and economic elements in evaluating the future of Hong Kong.

What is at stake is the existence in the twenty-first century of a nineteenth century mode of economic organization: a reliance on the competitive forces in a market economy. We have cast our explanation of Hong Kong's free market economy in both political and economic terms, but critics of this explanation, should they for the most part accept it, can counter with the assertion that Hong Kong is unique and that its economy is inapplicable or unsuitable to other societies for a combination of political, economic, international, and social reasons. They cite, in particular, the vestiges of colonial rule, arguing that

Hong Kong would not survive did it not serve so
economic interests of neighboring Communist
nd so on.

In this final chapter I want to bring before you some projections into the future of Hong Kong. Unfortunately, my crystal ball only hints at political and social trends, and not at property, commodity or share prices. But I can identify the parameters that hold in store Hong Kong's economic and political future. And then I want to use my analysis of Hong Kong's political economy to address the general question of economic freedom, *viz.*, under what conditions (and why are these conditions so historically infrequent) do free port, free trade, competitive market economies rise and flourish? Although I am looking down the road toward a theory of economic freedom, I content myself with an exploration into several historical instances of free market economies that I have been able to identify and research in the past three years, to see what they all have in common. But first let us focus into our crystal ball and complete the saga of Hong Kong.

The Future of Hong Kong

We have couched our explanation of Hong Kong's competitive market economy in terms of (1) the fact of its almost complete resourcelessness, (2) the tripod of consents that comprise its political geography, and (3) its constitutional and administrative status, which allows its economic and budgetary policy-makers to maximize private and public efficiency.[1] A succession of Hong Kong Financial secretaries, who understand and preach the principles of market economics, have practiced fiscal responsibility and economic noninterventionism. This combination of circumstances continues to attract funds to Hong Kong for investment and safe keeping, and allows its residents to enjoy a rising standard of living.

Let us follow this framework as we dial our time machine forward to 1997. Apart from an excellent harbor, convenient location, and hard-working labor force, Hong Kong will remain an externally dependent economy; resourcelessness dictates a harsh economic reality in that the government can do little to cushion the colony against exogenous economic events except to let the automatic corrective mechanism quickly run its course. Despite a recent investment in medium and heavy industry, including the processing of raw materials, the great bulk of Hong Kong's raw materials and foodstuffs will still need to be imported. To pay this bill, Hong Kong will have to export manufactured goods or import capital. The greatest uncertainties for Hong Kong's future lie in its political and economic unknowns. International and local business executives considering large-scale investment with slower returns than the norm in Hong Kong may be increasingly concerned at their inability to project the colony's political and economic stability for the next ten to twenty years.

Take electricity, for example. China Light and Power will be reluctant to expand capacity if future prospects for profits are bleak; non-investment of this sort would, in turn, retard industrial expansion. Similarly in the public sector, expenditure on Hong Kong's new underground railway and a projected new airport by the mid-1980s requires recourse to international loan finance: when will international lenders regard Hong Kong as excessively risky even though the Hong Kong government fully guarantees these loans? We here confront the political question of obligations should Hong Kong cease to exist in its present form. At this point, it is reassuring to note that a ten-year bond issue, denominated in Hong Kong dollars, for the Mass Transit System and guaranteed by the Hong Kong Government, was brought to the market on a yield of 9.375 percent in 1976 and was successfully oversubscribed. These bonds traded in spring 1977 at a healthy premium over issue price, yielding under 8 percent. The success of this bond issue suggests that investors in Hong Kong are attracted to long-term, Hong Kong-denominated paper; evidently relationships with China could not be better.

To return to electricity. Optimists over the colony's future point to current plans of China Light and Power for new generators and a distribution system priced at HK$8.4 billion (US$1.8 billion) that includes four 350-megawatt turbine generators.[2] The *Wall Street Journal* of 20 January 1978, reports that two British concerns signed letters of intent to provide equipment to the value of US$193 million to commence these plans for expansion of electric power generation. This investment demonstrates the long-term faith in Hong Kong of the management of China Light and Power.

The banking system, on balance, now takes a decade-long horizon in the investment of its assets. Until recently, banks typically loaned only 50 percent of the purchase price of a flat at 14 percent interest, to be repaid in a few years. Now loans of 80 percent for ten to twelve year redemption periods at 10 percent rates of interest are common.[3]

Moreover, the Mass Transit Railway Corporation, a quasi-governmental body, has extended the initial plans for a HK$6 billion mass transit system to one expected to cost HK$11 billion. Although the operating cash flow is projected to become positive by 1982, the MTR Corporation will not repay all its debts until 1992.[4] Still, the Corporation has encountered no difficulty raising the required loan finance.

Some cause for worry is found in the uncontrollable

policies of Hong Kong's trading partners. On 3 December 1977, Hong Kong signed a new five-year textile agreement with the EEC in Brussels, the most restrictive export agreement Hong Kong has ever accepted. Estimates of job losses run to ten thousand and revenue losses to HK$50 million per year.[5] This growing world trend toward protectionism in Hong Kong's chief export markets is viewed with anxiety in top government circles.

I'd like to reassess the tripod of consents that is Hong Kong's unique political geography. Recall that the post-1949 Communist government in Peking has renounced the "unequal treaties" which constitute Britain's claim to sovereignty in Hong Kong, Kowloon, and the New Territories lease. In Peking's official pronouncements, 1997 has neither more nor less political significance than 1977, 1987, or 2017. The problem for Peking is how to keep the Hong Kong economy buoyant and profitable if the 1997 issue becomes acute. Despite Chinese insistence that calendar year 1997 bodes no special significance, some local and overseas investors may find Peking's silence disquieting and begin to shift assets from the colony well before 1997. A massive capital outflow reduces Hong Kong's economic benefit to China, benefits which have thus far contradicted the implications of ideological dogma. I could find no official in Hong Kong last year who voiced serious cause for alarm, yet several suggest the start of a new decade will sharpen public concern over the future of colonial rule in Hong Kong. The fear is that Peking might regard it as politically infeasible to be seen actually giving a foreign government an extension of legal power over Chinese territory. Only if a massive outflow of capital begins, because investors regard the future as too risky, are the Chinese likely to take a public stand. What factors will enter into Peking's calculus?

First is the loss of valuable foreign exchange Peking derives from trade with Hong Kong. Second, a Communist Hong Kong will lose its international textile quota agreements, and require feeding by China with free grain imports for which it now pays Peking in hard cash. Third, China today operates far-flung business enterprises in Southeast Asia in the form of Hong Kong incorporated companies that might be put out of business should Hong Kong cease to exist. Southeast Asian countries that allow Hong Kong incorporated companies to trade in their respective nations might not grant the same privileges to explicitly China-incorporated firms; these firms would thus lose their markets and trading opportunities. Fourth, Hong Kong remains a laboratory to study capitalist institutions, serves as a base for intelligence activities and is even an outlet to release dissatisfied overseas Chinese who returned to China in the 1950s and 1960s. Finally, China would have to absorb nearly five million Cantonese who disavow the mainland's life style and politics and who cannot even speak its language. (Shanghai, Hong Kong's pre-war counterpart, remained a source of political difficulty some twenty-eight years after takeover.) Moreover, China would have to re-educate and relocate decadent, bourgeois, Westernized Chinese.

But no one forecasts with precision events in China. China-watchers have not scored high marks in evaluating leadership and policy changes. Try as we may to reason through Peking's present and future rationale for its tolerance (dare we say encouragement) of Hong Kong, this element of political risk will always remain. The best we can do is to watch Chinese utterances and behaviors and let the marketplace evaluate the economic risk.

It is interesting that China is building up its own investment stake in Hong Kong at an increasing rate since 1970. China is pouring money into the development of Tsing Yi

Island, one of the few remaining centrally located areas for industrial expansion. First came news of Peking's plans to build a HK$50 million machinery factory on the island. The latest project is the construction of a HK$100 million ship-repairing facility on the west of the island, for which a China-based firm is negotiating the purchase of 1.2 million square feet of land. It must strike some chord of curiosity that China is explicitly purchasing from the colonial government the use of its own claimed territory for a ship repair facility. Peking does not demand free use of land in Hong Kong and willingly lives by the verdict of the marketplace in Hong Kong's capitalist economy.

The *London Times* estimates China's investments in Hong Kong at a conservative US$2 billion.[6] On 20 December 1977, the newly built headquarters of China Products Company, China's major department store outlet in Hong Kong, opened in Causeway Bay stocked with more than HK$100 million worth of merchandise. This addition brought to 101 the number of China Products emporiums in the colony. At an opening press conference of this headquarters, Mr. Chang Cheng, chairman of the state-controlled firm, reiterated five times that "China is not anxious to alter the status quo as far as Hong Kong is concerned."[7] Moreover, he said, Hong Kong would enjoy stable food prices from China in coming years.

In a speech before the Commonwealth Club of California on 17 March 1978, the Chief Secretary, Sir Denys Roberts, echoed these remarks. He observed that relations with China were better than ever, pointing both to the expansion of Chinese economic activity in the colony and the provision of food supplies from China at very reasonable prices, which helps keep down the cost of living. He repeated the Chinese position which states that, pending a settlement, the status quo should be maintained. Although this formula contains no date, he confidently stated that Hong Kong will remain in its present form so long as it suits China's interest.

Many students of Hong Kong regard China's policy towards Macao as a bellwether of its future intentions in Hong Kong. On this score, the future is also bright. China has just built a new reservoir at Ta Ching San near the coastal village of Heung Chow six miles northeast of Macao, in order to increase its water supply to Macao. When it becomes operational in 1978, its daily supply of 10 million gallons will increase fourfold the combined capacity of the two Chinese reservoirs now supplying Macao.[8] At present, China is enthusiastically supporting economic development in Portuguese-administered Macao.

To return to leg two of the tripod of consents—the United Kingdom government. I'm beginning to think that the initials H.K. signify happy kingdom, and U.K., unhappy kingdom. In my admittedly Hong Kong bias, I sometimes fear the decisions of London more than those of Peking. The Chinese are, after all, a rational and hard-working people. Recent months have seen increased intervention by London in the affairs of Hong Kong. I, for one, suspect that a socialist-thinking Labor government, which increasingly nationalizes and taxes British industry and labor, finds embarrassing the success of lowly taxed capitalist Hong Kong. Over ten years ago, residents of Hong Kong endured a standard of income and living about one-tenth that of the average Briton; today the ratio is one-half. Projecting the respective rises in standards of living, the ratio will approach unity sometime in the 1980s; thereafter, Hong Kong residents will enjoy a higher standard of living than Her Majesty's home subjects. Hong Kong's success built on an antiquated market economy discredits Britain's modern socialist but declining economy. As the British pound has fallen in value against the U.S. dollar in recent years, from a rate of $2.40 to about $1.80 in spring 1978, the Hong Kong dollar has correspondingly appreciated, in the same period, from HK$5.00 to about HK$4.61 against the U.S. dollar, a rise of nearly 8 percent. The lack of trade unions in Hong Kong must call into question their usefulness for British economic welfare. Finally, Britain must resent economic competition with its own dependent territory, both in home and overseas markets.

There is an increasing sense of unease in Hong Kong that many in Whitehall and Parliament regard as inherently immoral the colony's low tax system, even though the budget enjoyed its largest surplus ever last year. The colony is among the world's most credit-worthy governments even as the mother country increasingly suffers from the English disease and mounting debts.

Since the end of the war, the Colonial Office, and its successor, the Foreign and Commonwealth Office, have had little reason to interfere in the internal affairs of the colony, apart from periodic pressure from Lancashire about textile exports, about which Labor governments are typically more responsive than Conservative governments. Both the United Kingdom ministers and the Hong Kong administration were pleased with this practice of benign neglect. Hong Kong's effective autonomy in internal affairs has remained unimpaired through 1978, despite increasing pressure from Whitehall for more labor legislation, social welfare spending, and higher taxes.[9] A return to power of the Tories would not displease Hong Kong authorities who, noting Britain's dismal economic condition, resent her meddling in the colony's labor affairs, social programs, and fiscal and economic policies.

In the Fall of 1973 I interviewed a cross-section of Hong Kong's public officials. My analysis of these interviews and other materials disclosed political divisions within the government bureaucracy, which pitted the Finance Branch and its supporters in one corner, against the "policy" branches, the departments, and their spending allies in the other. It was my feeling then that the Governor better understood British social and economic thinking than the realities of an externally dependent resourceless economy; the recession of 1974, however, has evidently taught him greater appreciation for the market economy.[10]

When I returned to Hong Kong in December 1976, I found that internal administrative disputes had given way to a unified dislike and resistance against intrusions of Her Majesty's Labor Government into the territory's internal affairs. There were considerable fears in Hong Kong that Britain wished to impose a welfare state, with a concomitant increase in rates of taxation on business profits and salaries. Hong Kong officials, of all persuasions, are rightly irritated when the representatives of a demonstrably unsuccessful economy try to impose their profligate and interventionist policies on a fiscally responsible, autonomous territory. Restoration of Tory rule, under which Hong Kong would more likely go its own way, might reopen internal bureaucratic lines, but perhaps all have learned their lesson from the 1976 perturbations in United Kingdom-Hong Kong relations.[11]

We now come to the third and final leg in our tripod of consents. Fully 58 percent of the local population are Hong Kong born and raised, a proportion that rises each year. Historically, the local Chinese, who have no political tradition of democratic participation have shown little interest in Hong Kong politics. Among the young, however, a Hong Kong identity has been evolving. Many of the Hong Kong-born Chinese youth think little of mainland politics in any concrete or narrow sense. However, youth-inspired local political activity has been on the rise. Student agitation supported a decision to increase the use of Chinese in official business and it also helped motivate the establishment of the Independent Commission Against Corruption. It seems unlikely that the younger generation, born in Hong Kong and with no memories of the chaos of the Chinese civil war, will be as ready to treat government administrators with the same silent respect as their parents have done.

Hong Kong youth are today increasingly well-educated, articulate and self-confident, and thus perhaps less content to accept the present system of British rule in Hong Kong. I, for one, would discount a good deal of this youthful rhetoric since few, if any, select the alternatives of crossing the border at Lo Wu or migrating to Taiwan. They have greater economic and political liberty in Hong Kong than elsewhere and when pressed on this point will readily admit it. However, Hong Kong youth need not attain reunion with China or evolution of an independent Hong Kong; widespread visible discontent may weaken Hong Kong's magnetism for investment. Thus far, many of the educated have found prosperity and contentment in Hong Kong's market economy, or attractive positions overseas. Nor is Hong Kong's labor force quick to follow the guidance of youthful radicals. One might say that the stoic and cooperative manner in which Chinese workers accepted wage cuts and shared reductions in workdays in the 1974 recession indicates that the prop of local consent is firmly in place.

The art of futurology dictates that we look into administrative evolution within the Hong Kong government itself. I have earlier remarked on the dramatic size and growth of Hong Kong's public service, whose total work force grew from an authorized number of about 17,500 in 1949 to about 86,000 in 1970, with a corresponding tenfold budget increase for personnel from less than HK$100 million to nearly HK$1000 million.[12] Growth in the public service continued to just under 100,000 for the 1973–74 fiscal year, but has since tapered off. For 1977–78, the public service is projected at 112,666 permanent positions at a cost of just under HK$2.5 billion.[13] Since 1960, manpower in the public service has grown by 125 percent and current expenditure (in nominal terms) has increased eight-fold. But for the moment, the size of the public service has stabilized, and expenditures on personal emoluments remain a prime target of scrutiny by the Financial Secretary, who is watchful to insure that the public sector does not outpace the growth of the private sector.

One suspects that a large and growing bureaucracy will find it difficult to keep their hands off the private affairs of Hong Kong's people. I have subtitled my first volume on Hong Kong *New Departures in Public Policy*, in which I set forth illustrations of programs that departed from the live-and-let-live approach of the Hong Kong government. I cited, as evidence, public assistance grants to the indigent and elderly, recently extended to include able-bodied unemployed males aged fifteen to fifty-five years; free primary education in all government schools (except the five English language schools); restriction of marriages to monogamous marriages; massive government efforts on collection and publication of official statistics, urban renewal, rent control, labor legislation, and pollution control. Since the publication of that volume in 1973, government has stepped up its spending and activities in the fields of housing, education, medical and health services, social welfare, labor legislation, country parks, and so forth. Let me cite just a few examples.

In keeping with the modified industrial land policy, government now grants land at special terms for industries which introduce new technology of value to the economy.

Land is typically made available to firms at below market rates.

Housing is the government's largest subsidized social program; it hopes to provide new flats for about 200,000 people a year, or a total of public housing for another 1.5 million people by 1984. To insure that this housing fulfills its social purpose, subsidized rents will be maintained, at the expense of those living in more expensive privately rented accommodation. As well, the Financial Secretary announced a new homeownership scheme to accommodate some 180,000 people by 1984–85 in his March 1977 Budget Address.

In education, the immediate priority is to provide nine years of education, for all and within the means of all, including three years of secondary education. This extends the current policy of providing compulsory free primary school education. Technical institutes and tertiary educational facilities are also undergoing rapid expansion. Enlarged medical and health services, more social welfare—especially the extension of public assistance to able-bodied men between fifteen and fifty-five, and additional labor legislation are all part of the Governor's recommendations for improving the quality of life in Hong Kong.

At the conclusion of his address to the opening session of the Legislative Council on 6 October 1976, the Governor, Sir Murray MacLehose, placed special emphasis on labor legislation and social welfare; he sought to insure that the people of Hong Kong could count on the essentials of life—education, medical services, housing, and necessary relief through social welfare. These public essentials are, he insisted, to be balanced by comparative economic freedom which is, in his view, compatible with social and commercial responsibility. "I am convinced that in the construction and preservation of this balance lie Hong Kong's best prospect for prosperity, social harmony and international respect."[14] Do you suppose the same words were spoken in early postwar Britain? What will stop the bureaucracy from becoming a special interest group dedicated to preserving and expanding the powers and activities of government?

Re-enter the Financial Secretary: Hong Kong's exposed and dependent economy, the constitutional preeminence of the Financial Secretary, and the persistence of a value-for-money ethic of financial administration should restrain the potentially harmful effects of growing government encroachment in the private sector. Short of some exogenous change beyond Hong Kong's power to control (e.g., embargoes on its exports, a shift in Chinese leadership and policy, an outbreak of madness in Parliament or Whitehall), the colony will, in all likelihood, remain the world's lowest-taxed, least regulated, most fiscally responsible government—an exemplary state of affairs. Even if Hong Kong steps up its government meddling for putative social and economic purposes, it will still hold out a comparative advantage for investment and productivity since nearly every other government is encroaching even more rapidly on private economic activity in their respective countries.

I've tried to write a recipe in which the ingredients of Hong Kong's future are clearly labelled, measured and blended in. The proof of the pudding, though, must be in the eating. If I earned more than the wages of an academic, I would not hesitate to invest my money in Hong Kong. But I did not come here to tell you how to make money in Hong Kong, only to tell you why Hong Kong is a haven of economic freedom.

Hong Kong is virtually unique in the practice of non-intervention and its reliance on the free play of market forces. Why is this so? How did it come about? Why does the Hong Kong government consciously restrict its economic activities and regulatory roles? Under what conditions will new regulations be imposed on Hong Kong's free-trade economy? And, to address a more general question, where and when in history can we find other examples of free-trading societies?

Some Preliminary Observations on Free-Trade Economies

A complete theoretical account of free-trade societies must incorporate those factors which bear upon the inception of free market arrangements, their subsequent development, stagnation, and/or final subordination to internal government control or external closure.[15] I have tried, in the past three years, to identify and research every contemporary and historical instance of a free-trade society I could find in order to chronicle its emergence, development, and, in most instances, ultimate decline. Comparative analysis of free-trade economies puts Hong Kong's uniqueness in better perspective: is its economic system a will-o-the-wisp or a recurrence of an oft-repeated historical instance of economic freedom?

A preliminary search for societies predicated upon free enterprise institutions disclosed a disappointingly small number. In addition to Hong Kong, those nations that today rely upon the free play of market forces are, by and large, the few remaining British colonies: Gibraltar, the Cayman Islands, and Bermuda. Each of these are fiscally and economically responsible in the Hong Kong sense. Monaco on the French coast is a similar political economy. Among other modern and developing nations, a few are experimenting with free trade export zones and tax haven arrangements, the development of which merit watching.

My first sustained effort at economic history has thus far

revealed relatively few instances of societies with free trading, competitive market economies. At this point I want to ask your indulgence on the difficult problem of defining economic freedom. Not even Hong Kong is perfect in this regard. What I have in mind is (1) a market economy, (2) a minimum of government economic activity or regulation, and (3) free port or free trade economic institutions. Economic freedom also goes hand-in-hand with low taxes and individual prosperity. Let me then list the fruits of my historical research and select, from this list, a sampling of stories to tell.

Historical Instances of Economic Freedom

I have not yet found the most convenient handle to group the many varied and interesting examples of free economies, so I will take the easy way out and try to list them chronologically, bearing in mind the fact of some overlapping and off-again, on-again instances of economic freedom.[16] I intend to exclude from this initial listing such contemporary mini-state tax havens as the Channel Islands, Cayman Islands, Monaco, Lichtenstein, Andorra, Ceuta, Melilla, Trieste, Canary Islands, Aden (now defunct), Labuan and Penang (now defunct), and so forth.

The Greek Island of Delos: 166–69 B.C.
Cyprus: under Ptolemaic rule, fourth century, fourteenth century
Alexandria, Egypt: seventh century
Champagne Fair Towns of France: twelfth–thirteenth centuries
Flemish City States of Belgium: eleventh–sixteenth centuries
Hanseatic League Cities: twelfth–nineteenth centuries:

Antwerp: fourteenth–sixteenth centuries
Livorno, Italy: 1593–1860
Genoa, Italy: seventeenth century
Tangier, British Colony: 1662–1683; as
International City: 1945–1957
Gibraltar: 1704–1978
Malta, British Colony: 1801–1811
Ionian Islands, Zante and Argostoli, British Colony:
1814–1862
Heligoland, British Colony: 1815–1890; within
Germany: 1890–1910
Minorca, British Colony: mid-eighteenth century
Singapore: 1819–1957
Great Britain: nineteenth century
United States: nineteenth century
New South Wales, Australia, British Colony:
1870–1900
Danzig: 1899–1940

Although the conclusions I reach today are, in part, conditioned by the few stories I select for telling, I am sticking to a chronological sample plan. I will tell you, in turn, about the Greek Island of Delos in the first century B.C., review with you several of the Flemish and French fair-towns and city-states of the middle ages, describe the Italian city of Livorno in the late sixteenth century, and talk about Gibraltar, Malta, and the Ionian Islands as instances of British Empire free port communities. From these illustrations, along with the foregoing analysis of Hong Kong, I will cull several general principles that underpin economic freedom.

Delos

Recall the first chapter in which I noted that Hong Kong's automatic corrective mechanism takes its cue from the fact of almost complete internal resourcelessness and hence heavy dependence on external trade. The case of Delos is similar.[17] In Bradford's words: "If one is tempted to ask why so small an island, *without any natural resources*, ever became what it did, then the answer can be given by any sailor. Delos is the last, and best, anchorage between Europe and Asia." (italics mine) It is literally the focal point of a Mediterranean seaman's world.

The greatness of Delos's fame contrasted sharply with the smallness of its area. The island is a rocky ridge of gneiss and granite, about three miles in length running north to south, and no more than 1,420 yards in its greatest width. It is dominated by 370-foot-high Mt. Cynthus, rising in the middle of the island. The coasts are rocky and mainly steep. The only shelter from the north wind is on the western side of the island where both the sacred harbor and the commerical harbor of ancient days were located. The commercial harbor was the focus of trade; towards the end of the second century B.C., it was lined with shops and warehouses. Below these warehouses, edging up the slope of Mt. Cynthus, are the remains of the town proper, from which stretched houses of the rich merchants towards the harbor of Skardana. The surface of the island is uniformly rocky and wind-swept.

It is not our purpose to discuss the religious importance of Delos, save to note that trade and temples typically go hand-in-hand. Merchants have always been eager to purchase security in both worlds. As early as the seventh century B.C., the island was the religious center for the Ionian League of islands. It was subsequently consecrated by Athens and remained, throughout its history, an important center of worship.

The island's first great period of prosperity emerged during the Persian wars, when its position in the center of the Aegean made it strategically and commercially important.

As a convenient mid-sea depot, Delos had already become by the third century B.C. a major corn market of the Mediterranean.

Let us pick up the politics of the Aegean at the turn of the second century B.C. No one Hellenistic kingdom was supreme among the Macedonians, Egypt, Syria, Pergamum, or Rhodes; rather, an unstable equilibrium prevailed. Piratical activity in the Aegean threatened this flourishing Delian commerce, thus Delos sought protection from Rhodes, which, as a great sea-power, was recognized by 188 B.C. as preeminent in the formation of a new island league. Rhodes had been rewarded for its assistance to Rome against the Syrian monarch Antiochus III, whose defeat signalled the end of Macedonian sea-power. Rome rewarded Rhodes with Lycia, part of Caria, and control of the eastern Mediterranean seas.

Rome's growing power and presence in the eastern Mediterranean, following its defeat of Hannibal, was to become the key to future economic events in Delos. Delos steadily grew as an important center for the buying and selling of wheat; economic buoyancy was reflected in the Island's new cosmopolitan community with arrivals from Southern Italy and the establishment of foreign owned banking firms.

Perseus of Macedon, Philip V's successor, set in motion the forces that were to bring Delos its greatest period of prosperity, however unintended. Perseus aroused Greek nationalism. He invited back into Macedonia absconding debtors, condemned exiles, and those who had fled on charges of treason. This notice was posted at Delos (and elsewhere), with promises of restoration of property to the exiles, to attract the attention of pilgrims; nor were the notices hidden from visiting Roman magistrates. A copy of the treaty between Perseus and the Boeotians was also set up in the temple of Apollo at Delos.

War subsequently broke out in 171 B.C. and Perseus was imprisoned by Rome three years later. Rome was now mistress of the Greek east. Only the sacrosanctness of Delos kept the island free from direct military hostilities.

To reward its ancient enemy, Athens, which had been faithful to Rome for the preceding twenty-five years, the Roman Senate surrendered Delos to the Athenians. Athens evicted the native resident Delians, who were henceforth to be known as Rheneans, and sent her own colonists to replace them.

Rhodes suffered most severely for its sympathetic support of Perseus. She not only lost most of her mainland possessions; worse, by the Roman Senate's creation of a free port on Delos, Rhodes lost her commercial supremacy; the Romans, in turn, enhanced their own commercial supremacy in the eastern Mediterranean. The Athenians, in their administration of Delos, were not permitted to levy harbor dues, or tolls on incoming and outgoing merchandise.

The free port of Delos literally ruined Rhodes. The annual harbor dues, on which its great commercial prosperity was based dropped precipitously from a million to 150,000 drachmae; Rhodes was no longer a commercial threat to Rome.

Commercial activity in Delos grew unchecked for the next fifty years. But in 88 B.C., a political storm gathered on the horizon, in the form of an outbreak of hostility between Rome and the king of Pontus, Mithridates Eupator. Mithridates undertook a series of military expeditions in defiance of Rome; running out of patience, the Roman governor of Asia provoked a war for which the Pontic king was the better prepared. With the cooperation of Greek provincials unhappy with Roman rule, Mithridates slew about 80,000 Italian traders and financiers. Greeks and pirates flocked to the standard of Mithridates, who was at once master of the Aegean.

Athens and Delos each had to choose between Rome and Pontus. Athens had already become anti-Roman, but the island did not follow suit. The separation of Athens and Delos is consonant with the fact that the Athenians in Delos were a minority compared to the large number of Italians stationed there. A majority of Delians were Roman, prospering by the operations of the capitalists of Rome, and thus held to its cause.

Athens sent an expedition to subjugate Delos, but the Roman general in charge of the defense of the island slaughtered the sleeping and drunken Athenian expedition. The Athenian failure was, however, retrieved by the fleet of Mithridates, which in the autumn of 88 B.C. sacked the island of Delos and massacred twenty thousand men.

Rome subsequently reoccupied the deserted island and restored Athenian administration several years later. After the Peace of Dardanus (85 B.C.) Italian families returned, the Agora of the Italians and much other war-time damage was restored; the business of merchants and shippers resumed.

The second Mithradatic war of 74 B.C. forestalled this brief return to prosperity. Forces in Mithridates' camp marked out Delos, a Roman naval center, for attack. The unhappy Delians were led away as slaves. On ultimate victory, the Romans reoccupied and repaired the town, but Delos was never again rich enough to have enemies.

The Roman attempt to restore the island to prosperity included the presence of a Jewish commercial settlement. Delos was still free from taxation, but the prosperity was gone forever. The reasons for it had ceased to exist with the shifting of trade-routes. Egypt and Syria now communicated directly with Rome and the developing civilization in Northern Greece was thrusting Byzantium and Thessalonica into greater commercial importance. Julius Caesar administered the final blow by planting a colony of

Romans on the old site of Corinth, to which the local trade of the Eastern Mediterranean at once flowed—Delos was abandoned even more rapidly than it was occupied.

The story of the free port of Delos has as chapter titles (1) its strategic location in the trade-routes of the eastern Mediterranean, (2) its complete internal resourcelessness, (3) its vital roles, successively, in the corn-trade, wheat-trade, and slave-trade, (4) its suitability for Rome as a free port to destroy Rhodian commercial prosperity and competition, (5) its repeated sacking, destruction, and depopulation during several Aegean and Greek-Roman wars, and (6) its ultimate economic decline with altered trade-routes. Roman commercial motives and the island's strategic trading location are the backdrop against which the Roman Senate declared Delos a free port. After 69 B.C., its location was no longer strategic.

Fairs and Fair Towns: Antwerp

The story of Antwerp, save its violent ending, is going to sound very much like that of Hong Kong, as a city where the free play of market forces vigorously prevailed.[18] Antwerp's inception as a market town can be traced back to the Champagne fairs of the twelfth and thirteenth centuries, and the permanent fair city of Bruges through the mid-fifteenth century. Its greatness followed directly upon the decline of Bruges in the late fifteenth century and prosperity abounded until its violent pillage by the Spanish in 1576.

Although European fairs date back to the seventh century (when King Dagobert founded the Fair of St. Denis), the Champagne fairs became important international markets only in the twelfth century. The great and growing exchanges between the two trading areas of Europe—the northern and the southern—were carried on at the frontier of the two areas, which happened to be situated in East-central France. Champagne's position at the intersection of ancient land routes leading from the Mediterranean to the North German frontier, and from Flanders to Central and Eastern France favored its development as an intermediary. Its ancient towns, Troyes, Langres, Rheims and Laon, were well placed at the focal points of the transcontinental traffic and therefore provided convenient meeting places for the merchants of Italy, Provence, Germany, and the Low Countries.

The fairs held in the towns of Champagne were especially favored by the policy of the counts of Champagne, who placed merchants under their guardianship, insured their protection and safe-conduct, and set up a network of roads. These feudal lords, the counts of Champagne, recognized that it paid to have foreign traders frequenting their towns. By 1180 the four towns held a series of six fairs staggered over the year; common to each was an opening week during which merchandise was exempt from taxes.

Convenience of location, guarantees of peace and security to foreign merchants, and impartial and disinterested laws attracted commerce, with its attendant prosperity. The fair town provided wide opportunity for free trade within a context of mercantile law and order. The capitalist fair differed from an ordinary market-day of feudal society by its special attraction to foreign merchants, by its emphasis on wholesale trade, and by its tendency toward free trade and legal equality among all merchants. Cox argues that "These towns demonstrated how it became possible for a society to prosper extraordinarily under conditions of free trade."[19] He particularly noted that the Jews, generally unpopular throughout Europe, prospered under the protection of the counts. "Free trade among all merchants was the rule, thus foreigners were able to trade freely with each other."[20] It was this very personal freedom which attracted foreign merchants.

About 1300, restrictions were put upon traditional commercial freedoms and the counts became involved in conflict with the French monarchs; as might be expected, the Champagne fairs declined.

Fairs in Flanders date from the 11th century, but their significance and expansion emerges in the course of the 13th century. The location of Bruges, in particular, placed it at the focus of all the major commercial currents of the time. Flanders, like Champagne, enjoyed a fortunate political beginning; early in the 10th century the feudal fiefs and baronies of the Southwestern Low Countries were assembled into a strong principality by the counts of Flanders. These counts granted freedom of movement to the merchants in their domain and guaranteed security in the possession of their goods, even as they resisted similar appeals from the burghers in Flemish towns to establish autonomous, capitalist towns. The counts especially opposed relinquishing power to self-governing councils in their cities. The local counts and dukes strongly encouraged freedom of commerce for the prosperity of their domains: this broad interest in removing restrictions to trade in a natural, entrepôt area led to competition among the towns for the favor of merchants.

I have said that Flanders, especially Bruges, was strategically located. The country lay on the coast of the North Sea, open to seaborne trade from every quarter; it was also cut across by navigable rivers and was provided with sheltered estuaries for harbors. By the middle of the 12th century, Flanders became the foremost industrial region in Northern Europe. Cloth, in particular, dominated commerce and drew to Flanders the merchants of the world; Bruges thus developed into another international mart, and eventual successor to the fairs of Champagne.

Political evolution within the towns of Flanders saw increasingly autonomous municipal governments take greater interest in the commercial activities of their burgesses, and impose urban regulations and legislation, of which the prime object was monopoly in the staple branches of foreign trade.[21] It was the power of the French and English kings and the German Hanse that prevented the whole of Europe from breaking up into a loose assembly of small economically independent areas, each dominated by a monopolistic town. Postan suggests that the monopolistic features of fourteenth and fifteenth century towns—the main theme of urban history—nearly drown out discussion of freer trade and freer towns, open towns with open trade. "...we know that the great fair towns of Northern Europe—Bruges, Antwerp, Bergen-op-Zoom—were free ports where strangers were allowed to enter and where trade with strangers was more or less unrestricted by any law, except perhaps the law of residence and brokerage."[22] But it was not these free towns which typified the new order; instead, highly regulated monopolies characterized countless small towns in Europe that sank into "well-regulated stupor."

Today's Hong Kong entrepreneur would have felt very comfortable in sixteenth century Antwerp, which had supplanted Bruges as the principal rendezvous point of foreign merchants in Europe. Since late in the thirteenth century, its lords, the dukes of Brabant, had granted privileges to visiting alien merchants. Annual fairs appeared by about 1320 in Antwerp as a vehicle to encourage woollen manufacture in Brabant. Scholars offer several reasons to explain the supplanting of Bruges by Antwerp: a more suitable waterway in the Scheldt than the Zwin, increasing commercial restrictions, and local political unrest in Bruges in the latter half of the fifteenth century. Merchants simply removed their houses and activities from

Bruges to Antwerp, where they could take advantage of greater commercial freedom. Antwerp sought not to develop significant foreign trade, but to become a neutral territory where overseas merchants could trade in perfect equality and under complete protection. Commercial freedom became the basis of the city's wealth.

In its rise to preeminence, Antwerp was relatively more favored by location than its competitors, including Bergen-op-Zoom. Several floods deepened the eastern Scheldt, thereby easing access to Antwerp; moreover, Antwerp was more favorably situated for land traffic. Finally, it enjoyed a lengthier commercial tradition than its chief competitors; the regime of the Antwerp fairs had granted their visitors the utmost commercial freedom.

Indeed, even the regulations of guilds in Antwerp operated very feebly. Antwerp, in Cox's words, "became the place above all others in the capitalist system where individualism has an opportunity to thrive. The able businessman had the widest scope for play of his ingenuity."[23] Freedom of thought, including religious freedom, found a home in this haven of economic freedom. Foreigners even occupied important positions in the administration of Antwerp.

The natives of Antwerp shared in this haven of commercial prosperity; they derived employment from the magnitude and multiplicity of commercial transactions and acted as brokers, warehouse agents, and bankers. They supplied a good deal of wharfage labor.

Antwerp in its heyday offered a greater variety of commodities on a more highly competitive basis than any other trade center in Europe. Merchandise, including gold and silver, poured in from Asia, England, the Baltic, and the Americas. Major commercial and financial activity became concentrated in the Bourse; Antwerp was the first city to allow such complete freedom to foreign merchants.

Businessmen came to buy and sell, borrow and lend, chiefly among themselves; the Bourse was a convenience for businessmen, not an institution specifically for the natives of the city. Futures markets arose with the standardization of commodities, and loans could be syndicated for foreign potentates.

Antwerp was a mercantile and financial entrepôt; it never became a great export industrial center. Skilled artisans produced chiefly for local consumption—the market for luxury goods was especially active. Local industry did not emerge, in part, because the great cloth-making and finishing towns of Flanders and Brabant poured their goods into the city, which offered no protection to local industry from overseas competition.

This tale of Antwerp, which I hope was as exciting as the tale of Hong Kong has been, ended on an unfortunate note with the Spanish invasion of the Netherlands and the pillage of Antwerp by Spanish soldiers in November 1576, which prompted merchants to leave the town. I sincerely hope history takes a different course in Hong Kong.

The elements of the political economy of freedom are now coming into sharper focus. Convenient location, liberal economic policies of local government imposed specifically to attract population and commerce, commercial freedom (with accompanying religious and intellectual freedom), few internal resources (apart from a hard-working and enterprising population), protection by the territorial lords, dramatic individual and community prosperity, and, finally, external destruction. Antwerp is today simply a major city in Belgium. I do, however, want to stress the element of human choice, *viz.*, that the counts of Champagne, Flanders, and Brabant all understood the benefits of economic freedom. Why the majority of their contemporaries did not, I cannot say. But their subjects paid the price.

Livorno

Livorno is a city on the Mediterranean coast in the Tuscany region of central Italy, located about sixty miles from Florence.[24] The town first assumed importance when, by a gift of the Countess Matilda of Tuscany to the Pisan church, it passed in 1103 into the power of Pisa. It was sold, successively, to the Visconti in 1399, to the Genoese in 1407 for 26,000 golden florins, and ultimately in 1421 for four times that price to the Florentines who sought an outlet to the sea. Livorno's subsequent economic evolution was tied up with the Medici family.

We may call Ferdinand I de' Medici, grand duke of Tuscany from 1587 to 1609, the true founder of Livorno; he built a small new harbor, gave asylum to refugees, and launched the town as a commercial and industrial center. His son and successor Cosimo II completed the work begun by the grandfather Cosimo I; the construction of the Medici harbor. The last of the Habsburg-Lorraine princes succeeding the Medicean dynasty, Leopold II (1797–1870), greatly enlarged the town, gave additional privileges to foreign merchants, and further improved the harbor. After the plebiscite of March 1860, Livorno, with the rest of Tuscany, joined Italy.

In recounting the tale of Antwerp, I emphasized a human element—the wisdom and farsightedness of the counts of Champagne, Flanders, and Brabant; I want even more to stress the salience of astute leadership in my exposition of the free port of Livorno.

When Cosimo de' Medici took over the city and fortress of Livorno, its economic importance was trivial and its population numbered barely 500. The main Pisan harbor, threatened by silting since 1421, ceased to be useful around 1540; thus, when Livorno came under the full authority of the Medicis, it was the only accessible harbor in

Tuscany. Economic activity in the port and in its hinterland responded to broad European economic currents, but particularly stagnated from 1573 to 1593 in the context of a general European economic crisis. Harbor activities dramatically increased from 1593 with the onset of free port status.

A number of important policy decisions illustrate the enlightened Medicean leadership. Livorno was opened to the Jews in 1548 after Portugal intensified its inquisition the previous year; a new customs law implemented in 1565 confirmed a liberal tariff policy; silk received a tax exemption; free storage of merchandise for one year was authorized with no fees charged for trans-shipment—all these measures preceded by thirty years the formal declaration of Livorno as a free port. Merchants were steadily attracted to Livorno's accessible harbor from 1565, culminating with Ferdinand's issuance of a great edict of tolerance in 1591 inviting all the merchants of the world to his city. By the seventeenth century Livorno became the major entrepôt of the Mediterranean, despite its having been a malaria-infested swamp but a few years earlier when it was a port of hard laboring shipping hands, new immigrants, a city without a fountain, doctor, teacher, or a priest, in which life was nasty, brutish, and especially short.

Livorno's community was truly cosmopolitan, attracting many nationalities by amnesty and promises of protection from civil wars and religious persecution. Livorno was a true melting pot in which small businessmen made their fortune. Maps of the eighteenth century show separate cemeteries for Turks, Armenians, Greeks, and Jews; the latter's merchants played a very important role from 1570. In his writing on Britain in the Western Mediterranean, Monk tells us that "Livorno being made a free port in 1593, it speedily grew into a great entrepôt of trade dominated by Dutch and English merchants."[25] It continued to prosper all through the seventeenth century distributing great quantities of English cloth and salt fish and providing English shipping with rich return cargoes.

This free port of Livorno prospered well into and through the eighteenth century in the context of a century of European rivalries and wars. Indeed, Livorno's commercial success probably contributed to the British decision to declare Gibraltar a free port in 1705.

The Early British Mediterranean Empire: Gibraltar, Malta, and the Ionian Islands

In many ways, Gibraltar resembles Hong Kong.[26] It is a free port, but its government subsidizes nearly half the territory's housing. Apart from its strategic physical shape and location at the entrance to the Mediterranean, it is a resourceless externally dependent economy, bordering a manifestly large and often hostile nation. Gibraltar is a dockyard economy that serves NATO and British forces, with tourism and construction comprising the balance of the economy. Attempts to develop light industry have thus far not been successful.

Even the Financial Secretary sounds out the same themes. As I read the budget debates of the early 1970s, I was struck by the Financial Secretary's emphasis on Gibraltar's externally dependent economy, his advice that government conservatively estimate its revenue, the need to bring public housing rents more in line with costs, and the importance of a surplus on recurrent account, both to augment reserves and finance capital expenditure. Only the numbers differed which reflects, of course, the facts of Gibraltar's smaller size, economy, and population.

The British captured Gibraltar in 1704. Its wartime value was beyond doubt; it would serve as a useful anchorage for ships bound in or out of the Mediterranean and as a control point for the strait. In peacetime, though, the garrison would be expensive to maintain.

These economic pressures induced Queen Anne to declare Gibraltar a free port—free of import and export imposts and duties—on 7 February 1705. Her Majesty-in-council ordered that directions be sent to the governor or commander-in-chief of Gibraltar not to permit any duty or impositions whatsoever to be laid or received from any ships or vessels, and that the port be free and open for all vessels, goods, wares, merchandise, and provisions. Thus Gibraltar became a free port, and the pattern of its trading was determined for centuries to come.

Gibraltar's previous masters always found it difficult to induce people to settle on the Rock, and the British response to this problem was to make Gibraltar a free port. Alien people of assorted origins came with the prospect of personal gain from economic freedom and British protection. Indeed, the military sought to limit the civilian population, a liability in wartime, by restricting the number of residences which could be built on the Rock.

To insure a steady supply of provisions for the fortress, and also to rival a prosperous Livorno, the only other free port in the Mediterranean, Queen Anne formally ratified Gibraltar's free port status by order-in-council in 1712. In the early eighteenth century, commerce was directed at provisioning the garrison; entrepôt activity emerged during the second half of the century. Commerce mushroomed after 1805, when Napoleon excluded British shipping from other European ports, complemented by an inflow of men and money during the Spanish revolution of 1808.

Throughout the 1800s Gibraltar assumed increasing importance as a base in the British world trading empire. It passed from military rule to a civil colony in 1815, then to a crown colony in 1830, although the prospering merchant community sought no representative institutions. Opening of the Suez Canal in 1869 further increased Gibraltar's commercial opportunities.

It is an interesting aside that one Spanish response to Gibraltar's prosperity and its hospitality for smugglers was to open Cadiz as a free port on 21 February 1829. Cadiz did not long remain a free port. It was suppressed in 1830, as its economic freedom was incompatible with the royal privileges and monopolies that made up the Spanish economy. Smuggling bothered the Spaniards of 1830 less than threats to their position posed by economic freedom.

Gibraltar is still a free port, though it depends on British grants-in-aid since Spain closed the border to labor and tourism at La Linea in 1966. What is the future of Gibraltar? In a referendum of September 1967, its residents voted to remain associated with Britain rather than return to Spain by a majority of 12,138 to 44. Gibraltarians enjoy real economic advantages in their free port protected by Britain, with democratic internal self-government in place of Spanish authoritarianism, and with low rates of taxation. As another aside, Spain herself is notably reluctant to surrender her own colonial duty-free enclaves of Ceuta and Melilla in Morocco.

The keys to Gibraltar's future lie in London and Madrid, as Hong Kong awaits words from London and Peking for resolution of its uncertain future. But this is not the place to analyze changing economic and political policies and attitudes. I simply state the remarkable similarities between the two free port British dependent territories. Mithridates undid Delos; the Spanish, Antwerp; and Livorno was peacefully incorporated into Italy; only time will tell what bells toll for Hong Kong and Gibraltar.

I want to close this run through history with a few words about Malta and the Ionian Islands. Before 1798, Malta had only an inconsequential trade. However, Britain acquired the island for its commercial and strategic value in the Western Mediterranean. "On 30 July 1801, Valetta was constituted a free port by order-in-council. Suddenly, in 1807, the island sprang into great prosperity, increased by every impediment placed by war on natural and legitimate commerce."[27] When Napoleon issued his decrees banning the import of British manufactured and colonial goods, Maltese trade and smuggling reached gigantic proportions. The war between England and Turkey interrupted direct shipments between them, making Malta the natural depot for this trade. Similarly, the American Embargo Act of 1808 increased the demand for Mediterranean produce previously furnished by America. Setbacks began in 1810, due successively to peace with Turkey, the repeal of the American Embargo Act, and the occupation of the Ionian Islands. As the Continental System of Napoleon crumbled after 1812, the prosperity of Malta vanished.

Between 1803 and 1813 Malta was administered by civil commissioners, whose powers were never explicitly defined, but whose style of administration balanced Maltese interests and British war needs. Sir Alexander Ball and Sir Hildebrand Oakes, successive civil commissioners, witnessed amazing economic progress. After all, Malta offered remarkable facilities for trade, and its central position suited the sailing capacity of small vessels from every part of the Mediterranean. In war time it was the ideal rendezvous of the licensed trader and smuggler. Between the opening of the free port in November 1801 and 1806, British exports to Malta rose to a quarter of a million pounds. By 1807–1808, exports rose in stages to three-quarters of a million and then three million pounds. For a brief period, Malta served as the central depot of Mediterranean trade under the general pressure of Napoleon's continental system and the American Embargo Act. But plague in 1813 was the ultimate economic disaster: it killed one in twenty of the population, suspended all commerce, and ruined the finances of the island.

Crown Colony government subsequently replaced civil commissioners in Malta, but prosperity could not recover since the outbreak of the Greek War of Independence almost entirely destroyed commerce between Malta and the Eastern Mediterranean. The only other periods of Maltese prosperity occurred during the Crimean War and when Britain increased spending on naval defenses in the late nineteenth century.

Economic freedom and concomitant prosperity in the Ionian Islands ran its course during fifty years of life as a British protectorate.[28] In the course of the Napoleonic Wars, the British took, for strategic reasons, the islands of Zante, Cephalonia, Ithaca, and Cerigo in October 1809, and Santa Mavra in April 1810. Lt.-General James Campbell was appointed as Civil and Military Governor. Two of the seven islands—Corfù and Paxo—remained in French hands. Paxo was taken in February 1814 and Napoleon's defeat transferred administration of Corfù into British hands on 24 June 1814. The convention of 5 November 1815 formed Corfù, Cephalonia, Zante, Santa Mavra, Ithaca, Cerigo, and Paxo into an ambiguous "United States of the Ionian Islands" under a British protectorate.

Ionian commerce had enjoyed the protection of the British flag from April 1812, the same year in which a Civil Commissioner on the Malta Model was appointed. To encourage Ionian trade, James Campbell declared Zante and Argostoli free ports to absorb some of the produce being locked out of Malta because of the plague.

Campbell's successor, Sir Thomas Maitland, further unbound traditional constraints that had made for an inefficient Ionian economy: he modified the law of entail, built roads, and simplified tariffs. Assisting him was the Greek War of Independence which, by destroying the mainland harvest, brought prosperity to the oil and currant trades in the Ionian Islands. Even as mainland Greeks suffered so severely in the war and its aftermath, great material progress was made in the Ionian Islands under British rule, which included completion of a fresh water aqueduct in Corfù, convalescent hospitals and lighthouses, extension of road building, libraries, and education. In Monk's words: "The Islands presented a striking contrast to the Hellenic Motherland, independent it is true, but impoverished, misgoverned and disorderly. It was in marked contrast, equally, to the conditions that had prevailed in the islands themselves before the British came."[29]

Commercial prosperity in the islands was no accident of Commissioner Maitland's good luck. One of the few authors with whom Maitland was acquainted was Adam Smith. Following Smith's dictums, he lowered as far as he could the Ionian tariffs, from which trade had suffered greatly, reduced the transit duty to one percent, and eliminated all direct taxes. Government drew its revenues chiefly from export duties on currants and olive oil.

International politics ended the British Protectorate of the Ionian Islands, and with its termination came a decline in the standard of living of its inhabitants. The treaty of union of the Islands with Greece was signed on 29 March 1864 and Britain thus ceded the Islands to Greece.

A Preliminary Thesis of Economic Freedom

We can now list a set of conditions that seem conducive to the emergence of free port, free trade political economies. To this point the list is not complete nor does it constitute a full set of necessary and sufficient conditions for economic freedom; I prefer, instead, that we treat these conditions as hypotheses to be explored in the course of further research.

1. The fact of economic resourcelessness necessitates a heavy dependence on external trade.
2. A strategic location at the crossroads of trade as an entrepôt between major empires and trading groups.
3. The decisions of political leaders to attract population and commerce by declaration of free port status of territories under their jurisdiction.
4. The existence of an ideology supportive of free trade and minimum government interference.
5. External military guarantees or limited military requirements.
6. Competent financial administration.

It is not surprising that free economies have a short half-life. Although Hong Kong still flourishes as a free port dependent territory of the Crown, our other examples have fared less well. The sword, the plague, external political events, and a growing tendency toward greater government regulation have extinguished prior instances of free economies. When a more comprehensive history of economic freedom is written, a task I want to finish in the next few years, it will probably reveal that economic freedom occurs but rarely and only in the presence of special circumstances. To conclude this volume, it suffices to say that Hong Kong is at one and the same time both unique and not unique. It is unique in the sense of being one of a handful of truly competitive economies. It is not unique in that such economies have cropped up intermittently throughout history.

Notes

The Evolution of a Free Society

1. The "automatic corrective mechanism" has been a recurring theme in the Financial Secretary's speeches since the early 1970s. His most explicit statements are found in *The 1977–78 Budget: Speech by the Financial Secretary, moving the Second Reading of the Appropriation Bill, 1977*, paragraphs 6–17, pp. 3–10 (mimeo); *The Hong Kong Economy: the Adjustment Process, 1973–77,* a paper read to the Hong Kong Society of Security Analysts by C. P. Haddon-Cave (Financial Secretary, Hong Kong) on 3 September 1976 (mimeo); *The Future for Hong Kong* by C. P. Haddon-Cave (Financial Secretary, Hong Kong), a paper read at *The Financial Times* Asian Business Conference held at the Hong Kong Convention Centre, 21–23 October 1975 (mimeo); *The Economy and the Public Finances: A Mid-Year Report*, a speech by the Financial Secretary at the annual dinner of the Hong Kong Society of Accountants on 8 August 1975 (mimeo); and *The Hong Kong Economy: the Later Phase of the Recovery* by C. P. Haddon-Cave (Acting Chief Secretary, Hong Kong), a paper read to the Hong Kong Management Association on 9 September 1977 (mimeo). I have tried in these opening pages to highlight in summary fashion the automatic corrective mechanism and its beneficial properties.

2. Current figures are found in the *Hong Kong Annual Reports*. My figures are taken from *Hong Kong 1976. Report for the Year 1975.* (Hong Kong Government Press, 1976), pp. 168–69, 180–81.

3. Ibid., p. 47.

4. This material is reproduced from one of my previous studies. See Alvin Rabushka, *The Changing Face of Hong Kong. New Departures in Public Policy* (Washington, D.C.: American Enterprise Institute and Stanford: Hoover Institution, 1973), pp. 6–7.

5. Quoted in *Hong Kong Annual Report*, 1957, pp. 2–3.

6. K. R. Chou, *The Hong Kong Economy: A Miracle of Growth* (Hong Kong: Academic Publications, 1966), p. 81.

7. Edward Szczepanik, *The Economic Growth of Hong Kong* (London: Oxford University Press for the Royal Institute of International Affairs, 1958), p. 183.

8. Two lengthier versions of this topic are found in Alvin Rabushka, *Value for Money. The Hong Kong Budgetary Process* (Stanford: Hoover Institution Press, 1976), pp. 4–11, and Norman J. Miners, *The Government and Politics of Hong Kong* (Hong Kong: Oxford University Press, 1975), pp. 1–46.

Politics and Economic Freedom

1. This summary of features is taken from the author's more intensive investigation of the politics of the budgetary process in Hong Kong. See Alvin Rabushka, *Value for Money. The Hong Kong Budgetary Process* (Stanford: Hoover Institution Press, 1976), pp. 3–4.

2. See Alvin Rabushka, *The Changing Face of Hong Kong* (Washington, D.C.: American Enterprise Institute, and Stanford: Hoover Institution, 1973), p. 11.

3. See *Value for Money*, pp. 14–22; *Rules and Regulations for Her Majesty's Colonial Service* (London: Her Majesty's Stationery Office, 1843), Chapter VII, pp. 52–60, Chapter IX, pp. 70–75, and Chapter X, pp. 87–88; and, *Colonial Regulations Being Regulations for Her Majesty's Colonial Service*, Part II–Public Business (London: Her Majesty's Stationery Office, 1951. Reprinted 1961), pp. 22–25.

4. *Regulations of the Hong Kong Government. Volume 3: Financial and Accounting Regulations* (Hong Kong: Government Printer, 1968).

5. This material is condensed from *Value for Money*, Chapter 3, pp. 38–82. See also Norman J. Miners, *The Government and Politics of Hong Kong* (Hong Kong: Oxford University Press, 1975), Part II, pp. 49–180. For a thorough, but perhaps unduly biased, treatment of the constitutional basis of Hong Kong government, see John Rear, "The Law of the Constitution," in Keith Hopkins, ed., *Hong Kong: The Industrial Colony* (Hong Kong: Oxford University Press, 1971), pp. 339–415). For the full extent of Rear's bias, see his other essay, "One Brand of Politics," in the same volume.

6. *Letters Patent and Royal Instructions to the Governor of Hong Kong* (Hong Kong: Government Printer, 1972).

7. Peter B. Harris, "Political Science and Political Style: Europe, Af-

rica, and Asia (An Inaugural Lecture)," University of Hong Kong, *Supplement to the Gazette,* vol. 18, no. 3 (February 1971), p. 12.

8. *The Machinery of Government: A New Framework for Expanding Services* (Hong Kong: Government Printer, May 1973).

9. See *Value for Money*, pp. 52–70.

10. Ibid., pp. 83–116.

11. The Financial Secretary, *Hong Kong Hansard 1955*, p. 61.

12. Ibid., 1963, p. 134.

13. Ibid., 1965, p. 77.

14. Ibid., 1964, pp. 50–51.

15. Ibid., 1960, p. 51.

16. Ibid., 1971, p. 537.

17. Speech moving the Second Reading of the Appropriation Bill, 27 February 1974, p. 40 (mimeo).

18. Ibid., p. 60.

19. *Hong Kong Hansard 1961*, p. 47.

20. Ibid., 1966, pp. 216, 218.

21. Ibid., 1972, p. 247.

22. Ibid., 1963, p. 50.

23. Quoted in G. B. Endacott, *Government and People in Hong Kong. 1841–1962: A Constitutional History* (Hong Kong: Hong Kong University Press, 1964), pp. 20–21.

Doing Business in Hong Kong

1. The title of this lecture is taken from an American Chamber of Commerce in Hong Kong publication. See Michele Kay, *Doing Business in Hong Kong* (Hong Kong: South China Morning Post & Amcham Publications Limited, 1976). I have made substantial use of this helpful volume in preparing this section, as documented in the footnotes.

2. Ibid., pp. 147–156.

3. For a comprehensive view of Hong Kong taxation, with reference both to cases and sources, see E. Newton, *Hong Kong Taxation–A Taxpayer's Guide* (Hong Kong: Department of Extramural Studies, The Chinese University of Hong Kong, 1975).

4. In the course of the 1977 Appropriation Bill, the Financial Secretary announced that the exemption limit on Estate Duty should be raised from HK$300,000 to HK$400,000. See *The 1977–78 Budget: Speech by the Financial Secretary, moving the Second Reading of the Appropriation Bill, 1977*, paragraph 221, pp. 104–105 (mimeo).

5. This history is summarized from a speech entitled "Direct Taxation in Hong Kong," delivered by the Financial Secretary, Mr. C. P. Haddon-Cave, at the Hong Kong Rotary Club on 12 August 1975 (mimeo).

6. See Annex (8), "Report of the Third Inland Revenue Ordinance Review Committee," in *The 1977–78 Budget.*

7. Kay, *Doing Business in Hong Kong*, pp. 173–204, and *Hong Kong 1976. Report for the Year 1975*. (Hong Kong Government Press, 1975), pp. 35–42.

8. Richard Hughes, "Hong Kong: A Personal View," in *Hong Kong 1976*, p. 6.

9. Kay, *Doing Business in Hong Kong*, pp. 33–64.

10. Ibid., pp. 121–146.

11. Richard Hughes, *Borrowed Place–Borrowed Time* (London: Deutsch, 1968).

12. Tom Pickens, "Hong Kong Business Opportunities," *Passages, Northwest Orient's Inflight Magazine,* December 1974, p. 9.

Is Hong Kong Unique? Its Future and Some General Observations about Economic Freedom

1. In light of the inscribed expiration date of the New Territories lease, the future of Hong Kong has been a topic of popular interest among journalists and academics. *The Asia Magazine* of 7 October 1973 ran a lengthy story entitled "This Incredible Relic. For the Crown Colony of Hong Kong, Time May Be Running Out," by John B. Koffend. The cover of that issue carried the title "Hong Kong: After 1997, What?" More recently, the *Far Eastern Economic Review* of 10 December 1976 featured on its cover page the story "The Future of Hong Kong," and carried lengthy analyses by Dick Wilson and David Bonavia.

One also typically finds in comprehensive political or economic analyses of the colony a section or chapter given over to questions of the lease and other political developments that make Hong Kong's future uncertain. My reading of the future reflects both similarities and differences with the foregoing essays; it is hopefully not too strongly colored by my admiration of Hong Kong's individualistic free society and my unwillingness to accept the possibility that Britain might one day hand over Hong Kong to the mercies of Peking.

2. *Far Eastern Economic Review*, January 20, 1978, p. 21.

3. Ibid., July 29, 1977.

4. *The Banker*, September 1977, pp. 51–55.

5. *South China Morning Post*, December 4, 1977.

6. *London Times*, January 31, 1978, p. 19.

7. *South China Morning Post*, December 19, 1977, p. 1.

8. Ibid., October 28, 1977, p. 11.

9. Address by H. E. The Governor, Sir Murray MacLehose, at the opening session of the Legislative Council on October 6, 1976 (Hong Kong: Government Printer, 1976).

10. The Governor, Sir Murray MacLehose, has written a foreword to the annual "Focus on Hong Kong" of the *Far Eastern Economic Review* of 18 March 1977. He writes, on page 37: "Looking back, I am impressed by the recession's brutal demonstration of the total dependence of the prosperity of Hong Kong on external markets. If their demand falls, Hong Kong's domestic market provides no cushion for its industry, and though the Government must always do what it can, it can provide no substitute for export orders, and its actions can only mitigate at the margins."

11. See Derek Davies, "A Touch of the Jitters," *Far Eastern Economic Review*, 18 March 1977, pp. 38–39.

12. Alvin Rabushka, *The Changing Face of Hong Kong* (Washington, D.C.: American Enterprise Institute and Stanford: Hoover Institution, 1973), p. 20.

13. *Supporting Financial Statements and Statistical Appendices for the Estimates of Revenue and Draft Estimates of Expenditure 1977–78* (Hong Kong: Government Printer, 1977), p. 842.

14. Address by the Governor at the Legislative Council, 6 October 1976, p. 14.

15. This research has been funded, in part, by a grant from the Law & Economics Center of the University of Miami. I thank Henry G. Manne, Director of the Center, for this support, but claim sole responsibility for the contents of this section.

16. Within the next few years I expect to complete a comprehensive history of free trading economies with supporting bibliography. I have chosen this lecture series to present the first set of preliminary findings and will accordingly cite only the key sources used in the preparation of this section.

17. This discussion of the free port of Delos rests on three main sources, the most important being W. A. Laidlaw, *A History of Delos* (Oxford: Basil Blackwell, 1933). The other two sources consulted are Ernle Bradford, *The Companion Guide to the Greek Islands* (London: Collins, 1975), pp. 120–22, and Jean-Phillippe Levy, *The Economic*

Life of the Ancient World (Chicago and London: The University of Chicago Press, 1967), p. 37.

18. I have pieced together this discussion of fairs and fair towns from Michael Postan, "The Trade of Medieval Europe: the North," in M. Postan and E. E. Rich (eds.), *The Cambridge Economic History of Europe, Volume II. Trade and Industry in the Middle Ages* (Cambridge: Cambridge University Press, 1952), pp. 119–256; Oliver C. Cox, *The Foundations of Capitalism* (New York: Philosophical Library, 1959), pp. 262–273; Robert-Henri Bautier, *The Economic Development of Medieval Europe* (Harcourt Brace Jovanovich, Inc., 1971), pp. 110–119; and, J. A. Van Houtte, "The Rise and Decline of the Market of Bruges," *The Economic History Review* 19, No. 1 (April 1966): 29–47.

19. Cox, *Foundations of Capitalism*, p. 264.

20. Ibid., p. 265.

21. Henri Pirenne, *Early Democracies in the Low Countries. Urban Society and Political Conflict in the Middle Ages and the Renaissance* (New York: W. W. Norton, 1971).

22. Postan, "The Trade of Medieval Europe," p. 222.

23. Cox, p. 268.

24. A comprehensive history of economic life in Livorno in the 16th and 17th centuries is found in Fernand Braudel and Ruggiero Romano, *Navires et Marchandises a l'entrée du Port de Livourne (1547–1611)* (Paris: Librairie Armand Colin, 1951).

25. W. F. Monk, *Britain in the Western Mediterranean* (London: Hutchison, 1953), p. 14.

26. I spent two weeks in May 1975 interviewing past and present officials of the Gibraltar government and searching through a number of historical sources in Gibraltar's Garrison Library. Two general and readily obtainable accounts of Gibraltar's history are John D. Stewart, *Gibraltar. The Keystone* (Boston: Houghton Mifflin, 1967) and Ernle Bradford, *Gibraltar. The History of a Fortress* (London: Rupert Hart-Davis, 1971).

Less readily available titles carefully preserved in the Garrison Gibraltariana collection provide vivid descriptions of commercial prosperity in 19th and 20th century Gibraltar: R. A. Preston, "Gibraltar, Colony and Fortress," *The Canadian Historical Review* (December 1946): 402–423; Gardiner, General Sir Robert, *Report on Gibraltar Considered as a Fortress and a Colony (1856)* (no publication information); H. W. Howes, *The Gibraltarian, The Origin and Development of the Population of Gibraltar from 1704* (Colombo: The City Press, 1951); José Plá,

Gibraltar (London: Hollis & Carter, 1955); and Captain Sayer, *The History of Gibraltar* (London: Saunders, Otley and Co., 1862).

27. C. Willis Dixon, *The Colonial Administrations of Sir Thomas Maitland* (London: Longmans, Green and Co., 1939), p. 159. The general summary of Maltese economic history is taken from the foregoing book and Monk, *Britain in the Western Mediterranean*, pp. 108–111, 170–173.

28. Three useful sources on British rule in the Ionian Islands are William Miller, *The Ottoman Empire and Its Successors, 1801–1927*, (Cambridge: Cambridge University Press, 1927), pp. 4–5, 40–45, 58–61, 120–124, 290–291; Monk, *Britain in the Western Mediterranean*, pp. 116–118, 161–162; and, Dixon, *The Colonial Administrations of Sir Thomas Maitland*, pp. 120–127, 179–191, 210–211, 229–239.

29. Monk, p. 162.